Amphibian Biology

Amphibian Biology

Edited by

Harold Heatwole
and
John W. Wilkinson

Volume 11

Status of Conservation and Decline of Amphibians:
Eastern Hemisphere

Part 3
WESTERN EUROPE

Published by Pelagic Publishing
www.pelagicpublishing.com
PO Box 725, Exeter, EX1 9QU

Amphibian Biology, Volume 11 Part 3
Status of Conservation and Decline of Amphibians: Eastern Hemisphere

ISBN 978-1-907807-52-7 (Pbk)
eISBN 978-1-907807-56-5 (ePub)
eISBN 978-1-907807-57-2 (PDF)
eISBN 978-1-907807-58-9 (Mobi)

British Library Cataloguing in Publication Data
A catalogue record for this book is available from the British Library.

Cover image: A pair of *Rana iberica* in amplexus. This species is regarded as of Special Concern and is the subject of a captive-breeding programme at the Peñalara Breeding Centre in Spain, aimed at recovering populations affected by chytridiomycosis and the introduction of alien species. Photograph by Cesar Ayres.

Table of contents of volume 11, Amphibian Biology: Eastern Hemisphere, Part 3 (Western Europe)

CONTRIBUTORS TO PART 3 (WESTERN EUROPE)

EDITORS

Heatwole, Harold, Department of Biology, North Carolina State University, Raleigh, Nc 27695-7617, U. S. A.
harold_heatwole@ncsu.edu
Wilkinson, John W., Amphibian and Reptile Conservation, 655A Christchurch Road, Boscombe, Bournemouth, BH1 4AP, Dorset, U.K.
johnw.wilkinson@arc-trust.org

AUTHORS

Belgium:

Bauwens, Dirk, Instituut voor Natuur-En Bosonderzoek, Kliniekstraat 25, 1070 Brussels, Belgium
dirk.bauwens@inbo.be
Louette, Gerald, Instituut voor Natuur-En Bosonderzoek, Kliniekstraat 25, 1070 Brussels, Belgium
gerald.louette@inbo.be

Britain:

Griffiths, Richard A., School of Anthropology and Conservation, University of Kent, Canterbury, Kent, CT2 7NR, U.K.
r.a.griffiths@kent.ac.uk
Wilkinson, John W., Amphibian and Reptile Conservation, 655A Christchurch Road, Boscombe, Bournemouth, BH1 4AP, Dorset, U.K.
johnw.wilkinson@arc-trust.org

Diseases:

Anderson, Lucy G., Faculty of Biological Sciences, University of Leeds, West Yorkshire LS2 9T, U.K.
anderson.lucyg@gmail.com
Bielby, Jon, Institute of Zoology, Zoological Society of London, Regent's Park, London, NW1 4RY, U.K.
jon.bielby@ioz.ac.uk
Bosch, Jaime, Departamento De Biología Evolutiva y Diversidad, Museo Nacional de Ciencias Naturales, CSIC, José Gutierrez Abascal 2, 28006 Madrid, Spain
bosch@mncn.csic.es
Cunningham, Andrew A., Institute of Zoology, Zoological Society of London, Regent's Park, London NW1 4RY, U.K.
a.cunningham@ioz.ac.uk
Fisher, Matthew C., School of Public Health, Imperial College, St. Mary's Campus Norfolk Place, London W2 1NY, U.K.
matthew.fisher@imperial.ac.uk

Diseases (continued):

Garner, Trenton W. J., Institute of Zoology, Zoological Society of London, Regent's Park, London, NW1 4RY, U.K.
trent.garner@ioz.ac.uk

Henk, Daniel A., School of Public Health, Medical School, Imperial College, St. Mary's Campus, Norfolk Place, London W2 1NY, U.K.
d. henk@imperial.ac.uk

Meredith, Anna, Royal (Dick) School of Veterinary Studies, Easter Bush Campus, Midlothian, EH25 9RG, U.K.
anna.meredith@ed.ac.uk

Martel, An, Department of Pathology, Bacteriology and Poultry Diseases, Ghent University, Salisbury 133, Merelbeke 9820, Belgium
an.martel@ugent.be

Pasmans, Frank, Laboratory of Veterinary Bacteriology and Mycology, Faculty of Veterinary Medicine, Ghent University, Salisbury 133, Merelbeke 9820, Belgium
frank.pasmans@ugent.be

France:

Vacher, Jean-Pierre, BUFO, Musée d'Histoire naturelle et d'Ethnographie, 11 rue de Turenne, 68000 Colmar, France
jean-pierre.vacher@edu.mnhn.fr

Miaud, Claude, Université of Savoie, UMR CNRS 5553 Laboratory of Alpine Ecology, 73 376 Le Bourget du Lac, France
claude.miaud@univ-savoie.fr

Ireland:

Marnell, Ferdia, Head of Animal Ecology, National Parks and Wildlife Service, Department of Environment, Heritage and Local Government, 7 Ely Place, Dublin 2, Ireland
ferdia_marnell@environ.ie

Netherlands:

Stumpel, Anton H. P., Reptile, Amphibian and Fish Conservation Netherlands (RAVON), Postbus 1413, 6501 BK Nijmegen, The Netherlands
sturie@planet.nl

Portugal:

Crespo, Eduardo, Centro de Biologia Ambiental, Departamento de Biologia Animal, Faculdade de Ciências de Lisboa, Bloco C2, Piso 2, Campo Grande 1749-016 Lisboa, Portugal
ejcrespo@fc.ul.pt

Domingues Castro, Maria José, Escola Universitária Vasco da Gama, Mosteiro de S. Jorge de Milréu, Estrada da Conraria, 3040-714 Castelo Viegas, Coimbra, Portugal and PROTERRA21, IPN - Instituto Pedro Nunes, Rua Pedro Nunes, 3030 - 199 Coimbra, Portugal
mariajcastro@gmail.com

Cruz, Maria João, Climate Change Impacts, Adaptation & Mitigation Research Group (CCIAM - SIM), Faculdade de Ciências de Lisboa, Bloco C1, Piso 4, Campo Grande 1749-016 Lisboa, Portugal
mjcruz@fc.ul.pt

Portugal (continued):

Oliveira, José Miguel, PROTERRA21, IPN - Instituto Pedro Nunes, Rua Pedro Nunes, 3030 - 199 Coimbra, Portugal
zmoliveira@gmail.com

Rebelo, Rui, Centro de Biologia Ambiental, Departamento de Biología Animal, Faculdade de Ciências de Lisboa, Bloco C2, Piso 2, Campo Grande 1749-016 Lisboa, Portugal
rmrebelo@fc.ul.pt

Teixeira, José, Centro de Investigação em Biodiversidade e Recursos Genéticos. Faculdade de Ciências da Universidade do Porto. Rua do Campo Alegre s/n. 4169-007 Porto, Portugal
jteixeira@mail.icav.up.pt

Spain:

Ayllon, Enrique, A.H.E, Apartado De Correos 191. 28911, Leganes (Madrid), Spain tesoreria@herpetologica.org

Ayres, Cesar, A.H.E-Conservation Issues, Apartado de Correos 191. 28911, Leganes (Madrid), Spain
conservacion@herpetologica.org, emysorbicularis@gmail.com

Bosch, Jaime, Museo Nacional de Ciencias Naturales, CSIC, José Gutierrez Abascal 2, 28006 Madrid, Spain
mcnbp3d@mncn.csic.es

Montori, Alberto, Universidad de Barcelona, Departamento de Biología Animal (Vertebrados), Universidad de Barcelona. Av. Diagonal, 645, E-08028 Barcelona, Spain
amontori@ub.edu

Ortiz-Santaliestra, Manuel, IREC-CSIC Universidad de Castilla-La Mancha, Ronda de Toledo, s/n Ciudad Real 13071, Spain
manuele.ortiz@uclm.es

Sancho, Vicente, Project LIFE-Anfibios Valencia), c/Felipe Valls, 40-7 esc-A, 46035 Benimàmet-Valencia, Spain
vicente.sancho@gmail.com

31 Infectious diseases that may threaten Europe's amphibians

Trenton W. J. Garner, An Martel, Jon Bielby, Jaime Bosch, Lucy G. Anderson, Anna Meredith, Andrew A. Cunningham, Matthew C. Fisher, Daniel A. Henk, and Frank Pasmans

Abbreviations and acronyms used in the text or references:

ATV	Ambystoma tigrinum *virus including Regina ranavirus*
BIV	*Bohle iridovirus*
DNA	*Deoxyribonucleic acid*
DRIP	Dermocystidium, *'rosette agent',* Ichthyophonus, *and* Psorospermium
ELISA	*Enzyme-linked immunosorbent assay*
FV3	*Frog Virus 3*
HE	*Haemotoxylin and eosin*
ID	*Infectious disease*
IUCN	*International Union for the Conservation of Nature*
OIE	*World Organization for Animal Health*
PCR	*Polymerase chain reaction*
PFU	*Plaque-forming units*
RaHV-1	*Ranid herpesvirus 1*
RCV-Z	Rana catesbeiana *virus Z*
RNA	*Ribonucleic acid*
SSRV	*Santee-Cooper ranavirus*
UK	*United Kingdom*
USA	*United States of America*

I. Introduction

Infectious diseases (IDs) are a *de rigueur* issue in amphibian conservation biology. The precipitous declines of Neotropical, Australian, and Californian amphibians due to chytridiomycosis (Berger *et al.* 1998; Lips *et al.* 2006, 2008; Vredenburg *et al.* 2010; Cheng *et al.* 2011) undoubtedly launched amphibian parasites onto the global stage, aided by a cast of viral, platyhelminth, protist, and other fungal infectious agents (Cunningham *et al.* 1996; Johnson *et al.* 1999; Green *et al.* 2002; Brunner *et al.* 2004; Raffel *et al.* 2008; Rohr *et al.* 2008b; Teacher *et al.* 2010). As a result, amphibian ID research has increased dramatically, with few tangible outputs relevant for applied conservation (Woodhams *et al.* 2011; but see Johnson *et al.* 2007), perhaps unsurprisingly, as characterizing the dynamics of disease and controlling IDs are thorny issues. When warranted, mitigating the effects of amphibian IDs will arguably be more complicated than in other terrestrial vertebrate hosts, due to cryptic hosts with complex life histories, large population sizes of hosts, and generalist parasites affecting speciose host communities (Lips *et al.* 2006, 2008; Johnson *et al.* 2008; Woodhams *et al.* 2011; Garner *et al.* 2012). Nevertheless, amphibians are declining at an alarming rate (Houlahan *et al.* 2000; McCallum 2007) and amphibian parasites are contributing factors in many of these declines (Skerratt *et al.* 2007; Crawford *et al.* 2010). We are obliged to respond to this threat, as human activities are factors in many novel amphibian host/parasite dynamics (Forson and Storfer 2006; Bosch *et al.* 2007; Fisher and Garner 2007; Picco and Collins 2008; Rohr *et al.* 2008b; St. Amour *et al.* 2008; Walker *et al.* 2008; Schloegel *et al.* 2009).

Although no record of an extinction of a European amphibian attributable to an ID has been published, evidence for mass mortalities, local declines, and extirpations of species certainly has been documented (Márquez *et al.* 1995; Cunningham *et al.* 1996; Bosch *et al.* 2001; Bosch and Martinez-Solano 2006; Ariel *et al.* 2009; Bielby *et al.* 2009; Balseiro *et al.* 2010; Teacher *et al.* 2010; Walker *et al.* 2010). Distributions of parasites have not reached saturation (*e.g.* Walker *et al.* 2010), so it is reasonable to expect expanding ranges and increased prevalence leading to increased deleterious effects on amphibian populations in Europe, as seen in other regions of the planet (Lips *et al.* 2008; Vredenburg *et al.* 2010). While Europe's amphibian biodiversity (88 species according to the latest IUCN Red List) is one or more orders of magnitude less than that of the Neotropics, Oceania, Africa, or Asia, it still harbours novelties, including many of the world's salamandrid genera, nearly all species of the Family Alytidae, the only proteid salamander found outside of North America and, until recently (Min *et al.* 2005), the only plethodontid salamanders known to occur outside the New World. Loss of European amphibian biodiversity may not have the potential to be as numerically significant as in other regions, but it would have taxonomic importance.

European amphibians are commonly affected by other threatening processes, including introduced and invasive species (Denoël *et al.* 2005; Bosch *et al.* 2006; Orizaola and Braña 2006; Segev *et al.* 2009; Ficetola *et al.* 2010), pollution (Orton and Routledge 2011), destruction and alteration of habitat (Ficetola and De Bernardi 2004; Ficetola *et al.* 2009) climatic change (Araújo *et al.* 2006; Reading 2007) and overharvesting/trade (Olgun *et al.* 2010). In response, conservation interventions based on sound research are being implemented across Europe and additional strategies for responding to threatening amphibian IDs must be integrated with these other conservation needs (Isaac *et al.* 2007). The impacts of parasites on amphibian populations are influenced by other threatening processes (Bosch *et al.* 2007; Rohr *et al.* 2008a,b; Garner *et al.* 2011) and because many other threats to European amphibians are ubiquitous, most conservation interventions developed in response to the threat of IDs will need to account for interactions with other threats (Isaac and Cowlishaw 2004; Blaustein *et al.* 2011). In some cases, IDs will be deemed relatively benign and resources may be better expended mitigating other threats (Teacher *et al.* 2010; Walker *et al.* 2010). Many published accounts of amphibian IDs are case studies of infections or reports of indirect

Fig. 31.1 Photograph of a UK common frog (*Rana temporaria*) exhibiting lesions suggestive of herpesvirosis. Photograph courtesy of Sarah Reed.

detection of parasites that assume presence equates to risk for amphibian hosts (Duffus 2010). To reach a balanced understanding of the level of risk, research on amphibian IDs must be focused, hypothesis-driven and multidisciplinary (Skerratt *et al.* 2009).

The list of amphibian IDs described in this chapter is limited to those parasites that research indicates are current or potential threats to European amphibian populations and species. The list should not be considered definitive, and should be improved and updated regularly by researchers and other stakeholders. Ideally it will shorten, but this seems unlikely. The state of knowledge regarding the conservation impacts of amphibian IDs on Europe's amphibian biodiversity is, for the most part, rudimentary. As knowledge grows, one can expect increasing evidence of the threat IDs pose for European frogs and salamanders. Hopefully the ability to mitigate threats from disease will grow accordingly.

II. Viruses

Viruses are one of two groups of microparasites responsible for the majority of recently emerging infectious diseases affecting humans, animals, and plants (Anderson *et al.* 2004; Jones *et al.* 2008). They are novel with respect to the other parasitic groups in this chapter due to their requirement for living host cells for replication. Viruses consist of a nucleic acid molecule encapsulated by a protein coat (and sometimes a lipid envelope) that cannot be individually visualized by simple light microscopy and therefore are difficult to detect. Additionally, manifestation of the disease may take unusual forms that could be accredited to non-infectious disease processes, or to nonviral parasites of amphibians (Marlow and Mizell 1972; Aguirre and Lutz 2004; Kang *et al.* 2008).

Viruses that infect amphibians include herpesviruses, adenoviruses, parvoviruses, ranaviruses and a calcivirus, most of which are not described in this chapter. Although reported in the veterinary literature, substantive evidence for the occurrence of many of these viral parasites in the wild or for either infecting or harming significant numbers of wild amphibians is lacking, as is quantitative data on any role they may play in amphibian population declines. Other pathogens, such as the amphibian erythrocytic viruses (a poorly studied group of viruses thought to be in the iridovirus family) (Hemingway *et al.* 2009) do occur in wild animals, but infections are not associated with any recognized impacts on survival or population dynamics.

In Europe, marsh frogs (*Pelophylax ridibundus*) have been shown experimentally to be capable of harbouring flaviviruses (Kostiukov *et al.* 1986) and to carry togavirus infections in Slovakia (Kozuch *et al.* 1978). Currently, only two groups of viruses are of conservation concern for Europe's amphibians: ranaviruses and herpesviruses (Johnson and Wellehan 2005).

A. Ranavirosis

Members of the genus *Ranavirus*, the cause of ranavirosis, are in the Iridoviridae, a family of double-stranded DNA viruses infecting a wide range of ectothermic hosts. Ranaviruses have a global distribution and infection of amphibians with ranavirus is notifiable to the World Organization for Animal Health (OIE) (Schloegel *et al.* 2010). Ranavirus strains capable of infecting amphibians and listed by the OIE include Frog Virus 3 (FV3) and all its synonyms; *Ambystoma tigrinum* virus (ATV), including Regina ranavirus; Bohle iridovirus (BIV), and *Rana catesbeiana* virus Z (RCV-Z). Santee-Cooper ranavirus (SSRV) and its synonyms, Singapore grouper iridovirus and *Testudo* iridivirus are also listed and considered as potential pathogens of amphibians. To date, all viruses that have been isolated from European amphibian hosts and characterized at the molecular level are either strains of, or closely related to, FV3 (Ahne *et al.* 1998; Hyatt *et al.* 2000; Balseiro *et al.* 2009; Holopainen *et al.* 2009).

Most of the understanding of the process of infection and viral replication is derived from studies of FV3 and it is unclear if, but assumed that, other ranaviruses exhibit similar life histories. Infection can be achieved by either naked or enveloped virions and both types exhibit broad host specificity (Goorha and Granoff 1979; Gendrault *et al.* 1981), although the mechanisms associated with entry into the host cell differ between these two types of virions (Braunwald *et al.* 1985). Post-infection, early viral gene transcription occurs in the nucleus, while later transcription is likely cytoplasmic (Goorha *et al.* 1978; Willis and Granoff 1978). RNA polymerization associated with immediate early expression of the viral gene is achieved using a combination of host polymerase and virus transcriptional transactivation factors (Chinchar *et al.* 2009). Transcription and translation of more delayed viral genes, including viral DNA polymerase, requires proteins translated during this process. Late viral gene expression also relies on viral DNA synthesized by viral DNA polymerase after transport into the nucleus of the host cell (Chinchar and Granoff 1984; 1986). Genome-sized viral DNA is methylated and forms concatamers after transport out of the nucleus of the host's cell, where virus assembly takes place at assembly sites in the cytoplasm. Experimental infections using *Xenopus* have demonstrated the development of the host's immunity to FV3 (Maniero *et al.* 2006; Morales and Robert 2007; Morales *et al.* 2010), but sequence data indicate that ranaviruses are effective at evading or modulating the host's immune responses (Essbauer *et al.* 2001; Jancovich *et al.* 2003; Morales *et al.* 2010). Chinchar *et al.* (2009) described ranaviral infection, viral replication, and evasion of the host's immunity in greater detail.

Infection by *Ranavirus* may result in one of four outcomes: (1) elimination of the virus without the development of clinical signs; (2) development of chronic, asymptomatic carrier animals; (3) development of an acute or peracute hemorrhagic disease which is invariably fatal, or; (4) development of chronic, cutaneous ulcerative disease that commonly results in death (Cunningham *et al.* 1996; 2007b; 2008; Robert *et al.* 2007). Both the probability of infection and its outcome are dose-dependent and route-dependent (Pearman *et al.* 2004; Cunningham *et al.* 2007b). Infection and viral proliferation may result in substantial inhibition of the host's ability to synthesize many molecules, including eukaryotic translational initiation factor 2α (Chinchar and Dholakia 1989). Function of this molecule for viral protein synthesis is presumably replaced by a viral homolog (vIF2α) that also influences virulence by inhibiting the host's antiviral protein kinase (Majji *et al.* 2006; Rothenberg *et al.* 2011). Infection with *Ranavirus* induces apoptosis that is responsible for much of the necrosis of tissue observed in diseased animals (Chinchar *et al.* 2009).

When evident, gross external lesions consist of erythema, oedema, and/or skin ulcerations, the latter mainly in more chronic presentations of the disease (Cunningham *et al.* 1996; 2008). The acute form or "hemorrhagic disease" as described for *Rana temporaria* presents as systemic petechial, ecchymotic, and frank hemorrhages. External lesions often are absent but when present can include cutaneous erythema and hemorrhagic, the vomiting of blood, and the passing of blood through the rectum (Cunningham *et al.* 1996; 2008). Microscopic examination often reveals pronounced focal necrosis in multiple organs, including the liver, spleen, kidney, and intestine (Cunningham *et al.* 1996; 2008). Intracytoplasmic inclusion bodies may be present in various cell types, particularly in mesenchymal and epithelial cells. Interestingly, in the more-chronic ulcerative presentation of the disease, the virus was not found in the spleen, an organ often recommended as a sample tissue for diagnosis (Cunningham *et al.* 2008).

The course of *Ranavirus* infection can be influenced by multiple factors and immunomodulation during metamorphosis, during hibernation, or shortly after hibernation, and can be associated with the development of clinical signs of disease (Essbauer *et al.* 2001; Jancovich *et al.* 2003; Morales *et al.* 2010; Robert 2010; Rothenberg *et al.* 2011). In contrast with events of mortality in the USA (Green *et al.* 2002), reports for Europe describe mass mortality in adult rather than in premetamorphic animals (Cunningham *et al.* 1996; 2007a, Ariel *et al.* 2009; Teacher *et al.* 2010; Kik *et al.* 2011) with two exceptions (Balseiro *et al.* 2010; Soares *et al.* 2002a,b). The development of clinical ranavirosis can be influenced by environmental metrics such as temperature (Rojas *et al.* 2005), but patterns observed in the laboratory do not explain why outbreaks seem more common in Europe during the warmer summer months (Gray *et al.* 2009). Ranaviruses in Europe appear to lack host specificity (Hyatt *et al.* 2000; Cunningham *et al.* 2007b; Balseiro *et al.* 2010) and pronounced differences in virulence exist among *Ranavirus* isolates as demonstrated in the common frog *Rana temporaria* (Cunningham *et al.* 2007b) and in North American host species (Schock *et al.* 2008; Hoverman *et al.* 2010). This suggests that host/*Ranavirus* coevolution might occur, reflected by the genetic composition of viruses affecting North American amphibians (Ridenhour and Storfer 2008). Teacher *et al.* (2009) described a pattern of host (*R. temporaria*) MHC class 1a genotypes that differed, based on the history of the disease in the population; tadpoles from genetically depauperate populations of Italian agile frogs (*Rana latastei*) were more susceptible to ranavirosis than were those from more genetically diverse populations (*Garner et al.* 2004; Pearman and Gomez 2005). Based on this evidence, it appears that European host species can adapt to the emergence of *Ranavirus*, but emergence should still be viewed as a substantial threat to European amphibian populations. Mass mortality events in Europe attributable to ranavirosis may involve thousands of animals (Ariel *et al.* 2009) and result in substantial and persistent population declines even in species capable of coevolution with *Ranavirus* (Teacher *et al.* 2009, 2010). It is probable that demographic effects can be exacerbated by other threatening processes (Forson and Storfer 2006; Kerby and Storfer 2009, Kerby *et al.* 2011).

The *post mortem* diagnosis of ranavirosis should be based on the presence of typical gross lesions (*e.g.* skin ulceration, systemic hemorrhagic) with positive results from at least one of the following

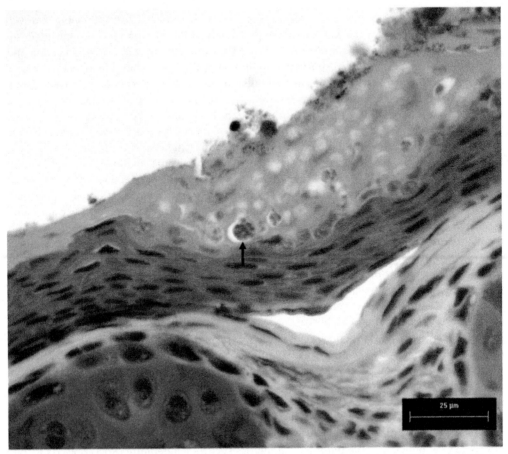

Fig. 31.2   Histological section of the epidermis of a natterjack toad (*Epidalea calamita*) lethally infected with *Batrachochytrium dendrobatidis*. Section is stained by haematoxylin and eosin and the *stratum corneum* exhibits thickening (hyperkeratosis). Thickened region has numerous developing zoosporangia and one obviously mature zoosporangium with a discharge tubule orientated towards the surface of the skin (arrow).

diagnostic tests: (1) *Ranavirus*-specific PCR of extractions of affected skin or internal organs; (2) the presence of viral antigen associated with typical microscopic lesions (*e.g.* focal necrosis) using immunohistochemistry; (3) culture of virus from diseased tissues, or; (4) visual detection of icosahedral virus particles using electron microscopy. Diagnosis in live animals is mostly accomplished using PCR analysis of skin swabs, or of clips from the tail or toes; clips tend to yield the most accurate results (Gray *et al.* 2009), but for European species the use of non-destructive sampling methods has not yielded accurate information. Serology is possible and has been used to detect seropositivity in cane toads (*Bufo marinus*) in Australia (Zupanovic *et al.* 1998) although there are no reports of it being used successfully in Europe. In the UK, attempts at serology, using the antibody-capture ELISA of Zupanovic *et al.* (1998) and virus-decoration technique, have failed to show seropositivity in wild *R. temporaria* from sites of endemic infection, and even experimentally-infected *R. temporaria* showed no evidence of seroconversion up to 30 days post-infection (A. Cunningham, unpublished observations). The duration of viraemia and its association (if any) with clinical disease has not yet been determined for *Ranavirus* infections. A blood smear taken during the viraemic phase and stained to determine the presence of intracytoplasmic inclusions in the erythrocytes may suggest ranavirosis but does not discriminate between ranaviruses and other erythrocytic iridoviruses [see Green (2001); Gray *et al.* (2009); Hemingway *et al.* (2009) for additional diagnostic information]. PCR of DNA extractions of blood samples could enable such differentiation, but have not yet been evaluated as a diagnostic technique.

In Europe, the first report of apparent ranavirosis in amphibians appears to have been in 1968, when mass mortality of *Rana esculenta* (now *Pelophylax* kl. *esculentus*) was reported in Croatia. A transmissible agent was cultured and shown to reproduce the systemic hemorrhagic disease when inoculated into frogs (Kunst and Valpotic 1968). Following this, Mişcalencu *et al.* (1981) reported finding ranavirus-like virions in the liver of *P.* kl. *esculentus* and Fijan *et al.* (1991) described the isolation of a *Ranavirus* from *P.* kl. *esculentus* with systemic hemorrhages and skin necrosis. It is unknown if any of these cases involved free-living, wild animals. Outbreaks of ranavirosis in wild amphibians in Europe have been reported for green frogs (*Pelophylax* sp.) in Denmark, Croatia, and the Netherlands (Fijan *et al.* 1991; Ariel *et al.* 2009; Kik *et al.* 2011); *R. temporaria*, common toads (*Bufo bufo*) and smooth newts (*Lissotriton vulgaris*) in the UK (Cunningham *et al.* 1996; Hyatt *et al.* 2000; Duffus and Cunningham 2010); common midwife toads (*Alytes obstetricans*), *B. bufo* and alpine newts (*Ichthyosaura alpestris*) in Spain (Balseiro *et al.* 2009; 2010; J. Bosch unpublished data); and marbled newts (*Triturus marmoratus*), Boscai's newt (*Lissotriton boscai*), *A. obstetricans* and the Iberian brown frog (*Rana iberica*) in Portugal (Froufe and Arntzen 1999; Soares *et al.* 2002a,b). Annually recurrent disease outbreaks in affected populations are reported for common frogs in the UK (Teacher *et al.* 2010). Additionally, ranaviruses have been isolated from green frogs in Italy (Holopainen *et al.* 2009) and from non-native, introduced North American bullfrogs in Belgium (*Lithobates catesbeianus*) (Sharifian-Fard *et al.* 2011).

Ranaviruses can be transmitted across damaged and intact epithelial surfaces through direct (skin contact, predation, cannibalism) or indirect (sediment, water) contact. It is thought that ranaviruses may also be transmitted vertically by infected ejaculates, eggs, or via shedding of viruses during reproduction, but this has yet to be shown definitively (Duffus *et al.* 2008). Environmental virions can persist for substantial periods of time, but cannot withstand even moderate periods of drying (Jancovich *et al.* 1997; Brunner *et al.* 2007; Gray *et al.* 2009). Experimental data indicates that infectious environmental concentrations of virions are in the range of 10^3 to 10^4 plaque-forming units (PFU) per ml (Rojas *et al.* 2005); however, Cunningham *et al.* (2007b) showed that frogs immersed for eight hours in an aqueous solution of approximately 22 PFU/mL could become infected. Regardless of the exact mechanism, an important goal of epidemiological studies is to determine the mode of transmission of infection, which may subsequently be used to plan suitable mitigation. The modes of transmission most commonly explored in attempts to understand transmission dynamics are density-dependent and frequency-dependent transmission. For the former, the rate of transmission occurs as a direct linear function of density of susceptible individuals; at higher densities transmission will increase linearly while at densities below a certain threshold, transmission will be unlikely to occur. The existence of a density-dependent threshold below which no new cases of infection occur is a key concept in epidemiology and in transmission of wildlife diseases. If transmission were solely density-dependent, populations should recover once the population size of susceptible individuals falls below the threshold level and, by extension, populations would not be driven to extinction. In contrast, with frequency-dependent transmission, the rate of contact between individuals is fixed and is not a function of population size; infectious contacts and transmission will occur at a similar level irrespective of host density. For this reason, frequency-dependent transmission does not have a density-threshold, which may greatly increase the likelihood of extinction of the host's population (De Castro and Bolker 2005) as new cases of infection will occur following reduction in density of hosts because of mortality in the population.

Greer *et al.* (2008) completed the only investigation of the transmission dynamics of *Ranavirus* using a single-host species/single life-history-stage host system in North America. They determined that transmission was not due to direct contact but through shedding of viruses even though the host's density played a strong role in determining rates both of transmission and of disease. Nevertheless, transmission dynamics were not best described by linear density-dependence. Instead, non-linear models (simple power or negative binomial functions) best explained the data and transmission dynamics such as these are unlikely to lead to extinction of the host through infection (Greer *et al.* 2008). These results could inform other simple host systems affected by ranavirosis, but most systems involve multiple hosts and life-history stages. Some of these can harbour asymptomatic infections or exhibit low mortality rates and act as competent reservoirs of infection (Brunner *et al.* 2004; Robert *et al.* 2007). It is probable that the patterns observed in simple host systems will not readily translate to these more complex communities (Duffus *et al.* 2008).

The introduction of infected amphibian and fish hosts is commonly accepted as a major source of ranaviral infection (Mao *et al* 1999; Ridenhour and Storfer 2008, Gray *et al.* 2009). In North America, redistribution of infected tiger salamander larvae used for fishing bait is arguably the most supported example of an introduced host increasing the distribution of ranavirus (Jancovich *et al.* 2005; Storfer *et al.* 2007; Picco and Collins 2008; Ridenhour and Storfer 2008). Quantitative evidence for any such mode of transmission is lacking for Europe but UK ranaviruses are genetically more similar to North American isolates than to continental European ones, suggesting trans-Atlantic incursion (Hyatt *et al.* 2000). Introduced North American bullfrogs and ornamental goldfish have been suggested as possible vectors (*e.g.* Duffus and Cunningham 2010; Sharifian-Fard *et al.* 2011), although these hosts are unlikely to be implicated in the emergence of ranavirosis at high elevations in continental Europe (Balseiro *et al.* 2009; 2010). Many alpine bodies of water are regularly stocked with game fish, which could act as reservoirs, along with water transported with fish and amphibian larvae that could 'hitchhike' with shipments of fish. The procedures proposed for limiting the anthropogenic spread of *Batrachochytrium dendrobatidis* (see below) should also serve to limit the human-assisted spread of ranaviruses; however, due to the proven ability of ranaviruses to persist in moist substrates, additional care should be taken with regards to the movement of fomites, such as aquatic plants and pond substrate.

B. Herpesvirosis

Herpesviruses (Family Herpesviridae) are double-stranded DNA viruses that cause clinical infections in a wide variety of vertebrates (Mettenleiter *et al.* 2008). While the herpesviruses of other vertebrates, especially those of mammals, are well-studied, little information is available on those affecting amphibians, with the exception of ranid herpesvirus 1 (RaHV-1), the causative agent of the Lucké frog kidney tumour in northern leopard frogs, *Rana pipiens* (McKinnell 1973; Davison *et al.* 1999). This virus is notable as the first reported oncovirus, *i.e.* the first virus known to cause malignancy (renal adenocarcinoma) (Lucké 1934, 1938). Uninfected adult leopard frogs exposed to RaHV-1 do not develop renal adenocarcinomas (McKinnell *et al.* 1989), but exposed eggs and developing embryos do and are presumed to become infected during the spring breeding period via virions shed by adult frogs (Green and Converse 2005). There are no reports of other species infected with RaHV-1. Experimental transmission to other species and the prevalence of silent infections in *R. pipiens* suggests that lack of detection should not be equated with lack of infection (McKinnell and DuPlantier 1970;).

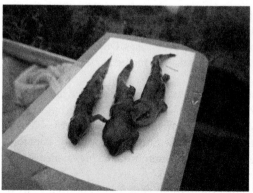

Fig. 31.3 Examples of pronounced *Amphibiocystidium* sp. infections of Scottish palmate newts. Signs of disease include cysts widely dispersed across the body, limbs, and tail, and previously undescribed oedema of the cranial region. Three newts in the left panel were found dead; the single newt in the right panel was alive when photographed.

Genomic studies indicate closer phylogenetic affinity of RaHV-1 to fish herpesviruses than to those from warm-blooded hosts, and methylation of host DNA is suggested as a component of the RaHV-1 infection and replication cycle (Mulcare 1969; Davison *et al.* 1999). Detection of virus in tumours is highly seasonal and temperature-dependent. Formation of viruses and viral inclusion bodies are detectable just before hibernation in winter, during post-hibernation and spring breeding, but are undetected during the warmer, summer months (McKinnell 1973). Conversely, growth of tumours, invasion, and metastasis occur primarily during the summer (McKinnell and Tarin 1984). Viral DNA can be detected in tumours during the warm season via PCR (Carlson *et al.* 1994). The prevalence of visible tumours varies in nature, but has not been reported in excess of 25% (McKinnell and John 1995). Prevalence declined sharply in affected *R. pipiens* populations in the 1970s and through the 1980s (Hunter *et al.* 1989), but may have increased in the 1990s (Green and Harshbarger 2001). As tumours are not always grossly visible, it is unlikely that the prevalence of visible tumours actually represents the true frequency of renal adenocarcinomas (Marlow and Mizell 1972).

Infection of amphibians by herpesvirus was first reported in Europe (Italy) in the mid-1990s in agile frogs (*Rana dalmatina*) (Bennati *et al.* 1994). The source of this outbreak, occurring in waves affecting first 35% then 80% of the population, remains unknown, but the consequences appeared rather benign. Frogs developed 1–3 discrete areas of epidermal hyperplasia, which resembled the "candle-wax" lesions described in fish with the herpesviral disease "carp pox". No mortality was reported during this event and co-occurring species (the Italian crested newt *Triturus carnifex*; green frogs *Pelophylax* sp.; and smooth newts *Lissotriton vulgaris*) were not visibly affected. Similar cases of cutaneous herpesviral disease have been seen in the UK (reports of multiple individuals of *R. temporaria*, affected at several sites since the early 1990s) (A. Cunningham and T. Garner, unpublished data; Fig. 1) and in Germany, where several cases have been reported in the European common spadefoot (*Pelobates fuscus*), *R. temporaria*, and *R. arvalis* (Mutschmann and Schneeweiss 2008). As in the Italian cases (and as is the case for "carp pox"), skin lesions were visible during the spring breeding period and invariably regressed over several weeks. It is not known if affected frogs exhibit lesions in subsequent years. Interestingly, five affected adult *P. fuscus* taken into captivity for further observations during the German emergence event all died some months later. *Post mortem* examinations revealed the presence of kidney tumours (adenocarcinomas) in all five animals (F. Pasmans unpublished data), findings consistent with reports of herpesvirosis-induced carcinoma in frogs in the USA (Lunger *et al.* 1965; Granoff 1983; Twedell 1989). Several *P. fuscus* collected in 2009 in northern France exhibited skin lesions that were highly suggestive of herpes-virosis and comparable to those described by Mutschmann and Schneeweiss (2008). Legislation prevented the sampling of these animals for the presence of herpesviruses. One *B. bufo* in Belgium exhibiting hyperpigmentation tested positive for herpesvirus using a molecular diagnostic test (F. Pasmans unpublished data).

Overall, herpesviruses may be rather widespread in Europe and appear to have the capacity to cause lethal carcinoma in at least one species (*P. fuscus*) of conservation concern. No information regarding population effects of herpesviruses for European amphibian species is at hand and, given the evidence available for North American species and the European common spadefoot, linking skin lesions to mortality rates will require more than data merely on the prevalence of skin lesions.

Diagnosis of herpesvirosis can be performed using PCR in combination with the occurrence of clinical signs (renal carcinoma and skin lesions) as well as by histological examination of skin lesions and suspect carcinomas (Green and Harshbarger 2001). Isolation of viruses can be difficult and success is very inconsistent; electron microscopy often results in many false negatives, so use of either should be in concert with the other listed diagnostics.

III. Bacteria

Bacteria are the other group of microparasites responsible for large numbers of recent emergences of infectious diseases in animals (Jones *et al.* 2008). They are prokaryotic single cells encapsulated by a phospholipid bilayer cell membrane. Fatty acids are common in bacterial cell membranes, whereas sterols are lacking. Cell walls contain peptidoglycans, and variation in content of these defines the two basic bacterial groups; Gram negative (thin peptidoglycan layer) and Gram positive (thick peptidoglycan layer). Bacterial agents, such as Mycobacteria or Chlamydiales, that are known to cause disease in captive amphibians are fastidious organisms and difficult to culture. In addition, typical mycobacterial granulomata as observed in many other vertebrate hosts are much less obvious in amphibians. Overall, both types of bacteria are easily missed during routine bacteriological examination.

Mycobacteria are well known pathogens of captive amphibians and can cause mass die-offs in captivity (*e.g.* Trott *et al.* 2004; Chai *et al.* 2006; Suykerbuyk *et al.* 2007; Sanchez-Morgado *et al.* 2009), but their role in disease of wild amphibians is unclear and understudied. Infections by Chlamydiales (*Chlamydophila psittaci, C. pneumoniae, C. abortus, Chlamydia suis* and 'Candidatus *Amphibiichlamydia salamandrae*') appear to be common in captive amphibians (Mutschmann 1998; Reed *et al.* 2000; Blumer *et al.* 2007; Martel *et al.* 2012) and have been associated with high mortality rates in the fire salamander, *Salamandra salamandra*, and other caudate amphibians (Martel *et al.* 2012). Infections occur in wild European amphibians, but some reports are of low prevalence and weak virulence. For example, Blumer *et al.* (2007) reported a prevalence of 2.5% of Chlamydiales in 126 *R. temporaria* but no clear link with disease. In contrast, A. Martel and F. Pasmans (unpublished data) describe the novel 'Candidatus *Amphibiichlamydia ranarum*' occurring in 71% of livers sampled from *R. catesbeianus* tadpoles in the Netherlands. A moribund *Alytes obstetricans* in Belgium infected with *Chlamydia* sp. and exhibiting atypical signs of disease in nature (apathy, skin depigmentation) died in captivity. No other causes of disease could be identified. Evidence of the zoonotic pathogen *Chlamydia pneumoniae* infecting wild amphibians suggests a potential public health risk (Blumer *et al.* 2007; Mitchell *et al.* 2010).

Disease, mass mortality, and declines of amphibians caused by 'red-leg' disease were at one time commonly attributed to *Aeromonas hydrophila*, a gram-negative heterotroph. Multiple reports of mass mortality in captive colonies of frogs at the turn of the century led researchers to experimentally investigate the cause of apparent disease. These early investigations satisfied Koch's Postulates and spawned both the name of the disease and the belief that *A. hydrophila* was the causative agent (Emerson and Norris 1905); however, the significance of these results to natural disease has been questioned (*e.g.* Gloroiso *et al.* 1974; Cunningham *et al.* 1996). Recent examples of mass mortality attributed to 'red-leg' and *A. hydrophila* include the extirpation of a population of mountain yellow-legged frogs (*Rana muscosa*) in the Sierra Nevadas of California and die-offs of boreal toads (*Bufo boreas*) in the mountains of Colorado (Bradford 1991; Carey 1993;). Subsequent research has since shown that the infectious disease responsible for these was most probably *Batrachochytrium dendrobatidis* (Muths *et al.* 2003; Vredenburg *et al.* 2010). Red leg syndrome in amphibians is now viewed as peracute or acute bacterial dermatosepticaemia that may be associated with one or more of a variety of gram-negative bacilli as well as some gram-positive bacteria. Several authors (*e.g.* Cunningham *et al.* 1996; Densmore and Green 2007) argued that historical overdiagnosis of red-leg syndrome epizootics is a serious issue because post-mortem and peri-mortem bacterial invasion is rapid in dead and moribund amphibians, and can mask the primary etiological agent. Furthermore, the majority of bacteria associated with bacterial dermatosepticaemia are environmentally ubiquitous and commonly isolated from healthy amphibians (e. g. Hird *et al.* 1981; Cunningham *et al.* 1996). The term 'red-leg' is a non-specific description of skin erythema with or without hemorrhages and in the UK it has been used when referring to cases of ranavirosis affecting *Rana temporaria* (Cunningham *et al.* 1996). Currently, there are no bacterial infections that have been shown to pose a significant conservation threat to European amphibian populations and most true bacterial infections, as opposed to *post-mortem* bacterial invasions, should be considered as secondary infections unless proven otherwise.

IV. Fungi

Fungi are highly diverse and ubiquitous eukaryotes with number of species estimated to be in the millions, with an average human breath containing 1–10 fungal spores (Frohlich-Nowoisky *et al.* 2009; Blackwell 2011). Fungi are saprophytic, parasitic, mutualistic, or often some context-dependent combination of the three (Stajich *et al.* 2009). Despite their ubiquity and diversity, fungi are only rarely pathogenic to vertebrates with only hundreds of species thought to infect them compared to hundreds of thousands of species infecting plants (Robert and Casadevall 2009). Infections by otherwise 'environmental' fungi commonly appear in response to weak immunity but, because of their flexibility, fungi are highly adaptable and often difficult to eliminate once they have become established (Garcia-Solache and Casadevall 2010). In terrestrial environments, Ascomycota and Basidiomycota are the most diverse and best-known lineages. Comparatively little is known about the fungal component of aquatic environments, and entirely new lineages of uncultured fungi have been discovered recently based on molecular data (Jones *et al.* 2011). Fungi feature multiple morphologies, developmental plans, and dispersal strategies. Their life cycles can be complex and may alternate between haploidy and diploidy as well as between multi-cellular hyphal stages and unicellular stages. Multicellular stages are composed of interconnected filaments called hyphae that can fuse together to form large net-like complexes that may also form complex macromorphological structures (*e.g.* mushrooms). Some fungi are known for producing copious numbers of airborne spores, and many produce long-lived and highly resistant spores. At least two lineages of fungi, the Blastocladiomycota and the Chytridiomycota, have flagellated, free-swimming aquatic zoospores (James *et al.* 2006). Fungi commonly employ asexual reproduction that enables them to rapidly invade new environments and produce clonal epidemics (Taylor *et al.* 1999).

Several fungal pathogens of amphibians are recognized in the veterinary literature but are not covered by this chapter. These include *Mucor amphibiorum* (experimentally linked to mortality in

Fig. 31.4   Cluster of *Amphibiocystidium* lesions on the cloaca of a male palmate newt. Lesions would be considered preulcerative based on the criteria of González-Hernández *et al.* (2010).

some European amphibian species) (Frank *et al* 1974; 1976) and *Basidiobolus ranarum* and several ascomycetes associated with chromomycosis. Infection with a member of the notable pathogenic genus *Cryptococcus* and signs of pulmonary cryptococcosis have been reported for *B. bufo* in Portugal (Seixas *et al.* 2008), but although heavily infected, detection was in a road-killed animal. For further information on fungal parasites of captive amphibians, the reader is referred to Taylor's chapter in Amphibian Medicine and Captive Husbandry (Taylor 2001) and to Berger *et al.* (2009).

A. Chytridiomycosis

The most widely studied of amphibian infectious diseases, chytridiomycosis, is notable for its propensity to cause amphibian declines and its implication in amphibian extinctions (Stuart *et al.* 2004; Wake and Vredenburg 2008; Lötters *et al.* 2009; Crawford *et al.* 2010; Heatwole 2013). The causative agent is the chytridiomycete fungus *Batrachochytrium dendrobatidis* (*Bd*), which like ranaviruses, is listed by the World Organization for Animal Health (OIE) as a notifiable pathogen (Schloegel *et al.*, 2010). Chytridiomyosis is characterized by the colonization of the keratinized layers (*stratum corneum*) of amphibian epidermis (Fig. 31.2) or of keratinized larval mouthparts of anurans (Berger 1998; Pessier *et al.* 1999). Infection occurs when flagellated motile zoospores invade keratinized cells (Longcore *et al.* 1999; Berger *et al.* 2005; Moss *et al.* 2008), where they mature into spore-forming zoosporangia and, in the case of metamorphosed animals, migrate to the skin's surface along with the epidermal cells they infect. During maturation, zoosporangia develop a discharge tube that opens to the surface of the skin, allowing the release of zoospores to the environment. Mature zoosporangia containing zoospores are also shed during sloughing of the skin (Berger *et al.* 2005). It is unclear how well *B. dendrobatidis* may persist and remain infectious outside an amphibian host (Johnson and Speare 2003; Rowley *et al.* 2007; Walker *et al.* 2007) and a substantial proportion of released zoospores may reinfect the original host (Briggs *et al.* 2010).

Clinical infection is associated with hyperplasia, hyperkeratosis and, in some species, excessive shedding of the epidermis (Berger *et al.* 1998; Berger *et al.* 2005). Anuran larvae may exhibit depigmentation and deformation of the mouthparts (Lips 1999), but not always. Physiological effects of infection include the disruption of the skin's osmoregulatory function, leading to dehydration, electrolyte imbalance, and death due to cardiac arrest (Berger *et al.* 1998; Voyles *et al.* 2007; 2009;

Carver *et al.* 2010; Marcum *et al.* 2010). Proteolytic enzymes that degrade components of the skin appear to play a major role in this process (Piotrowski *et al.* 2004; Symonds *et al.* 2008; Moss *et al.* 2010). The pathogenesis of the disease (invasion of the host's cells, intracellular proliferation, possible production of toxins), however, is still not fully understood (reviewed by Voyles *et al.* 2011), but the host's susceptibility varies strongly, both among and within host species (Woodhams *et al.* 2007; Walker *et al.* 2010). After infection, animals can eliminate the fungus, become persistent and asymptomatic carriers, or develop clinical signs of disease, which is generally associated with an increase in the fungal load (Vredenburg *et al.* 2010). Factors influencing susceptibility include hosts' innate immunity (Woodhams *et al.* 2007; Ribas *et al.* 2009; Lam *et al.* 2010), the composition of the skin microbiota (*e.g.* Brucker *et al.* 2008a; 2008b; Harris *et al.* 2006), adaptive immunity (Ramsey *et al.* 2010), and environmental factors such as temperature, humidity, and habitat alteration (Becker *et al.* 2012; Garner *et al.* 2011; Ribas *et al.* 2009; Andre *et al.* 2008; Bosch *et al.* 2007). Variation of virulence among isolates has also been described (*e.g.* Berger *et al.* 2005; Retallick and Miera 2007; Fisher *et al.* 2009b; Farrer *et al.* 2011).

A clear distinction should be made between detection of *B. dendrobatidis* DNA, evidence for infection with *B. dendrobatidis*, and clinical chytridiomycosis. The diagnosis for infection in living animals is typically accomplished according to the recommended standard protocol of Hyatt *et al.* (2007): skin swabs are the preferred sample and the qPCR described by Boyle *et al.* (2004) the preferred method of detection. Swab samples only detect pathogen DNA, though, and actual infection should be confirmed with histopathology or some other visualization of infection. In view of the relatively large number of asymptomatic carrier species, the diagnosis of clinical chytridiomycosis should consist of the combination of: (1) the presence of clinical signs such as disecdysis, apathy, or skin lesions; (2) the presence of *B. dendrobatidis*, preferably assessed using qPCR, allowing a quantitative estimate of fungal load; (3) demonstration of histopathological changes in the skin of infected animals using histology, including immunohistochemistry (Fig. 31.2), and (4) comprehensive *post mortem* examinations that exclude other causes of mortality, both infectious (viral, bacterial, fungal, macroparasitological) and non infectious (*e.g.* toxicological, environmental). Diagnostic methods for chytridiomycosis have been described in great detail by Berger *et al.* (2009): also refer to Green (2001). Additionally, every effort should be made to isolate and culture the fungus from infected hosts, as data on genotype are beginning to reveal previously undescribed variation of *B. dendrobatidis* that correlates with virulence (Farrer *et al.* 2011; see below) and pure isolates are required to generate fungal genotypes. It is hoped that simpler, multilocus diagnostics will eventually be available for characterizing fungal genotypes from archived and preserved amphibians.

Duffus and Cunningham (2010) reviewed the current distribution of *B. dendrobatidis* in Europe and European amphibians, which is being updated regularly as part of the European effort to map the distribution of the fungus (www.bd-maps.eu/). Infection, primarily determined through molecular diagnostics, has been demonstrated in 40 European species as of June 2011 (www.bd-maps.eu/). Direct evidence for lethal chytridiomycosis is reported or recorded for the common midwife toad (*A. obstetricans*) in Spain, Switzerland, France, Portugal, Germany, and Belgium and the Mallorcan midwife toad (*Alytes mulentensis*) in Spain (Bosch *et al.* 2001; Walker *et al.* 2008; Pasmans *et al.* 2010; Tobler and Schmidt 2010; J. Kielgast, personal communication); the Iberian midwife toad (*Alytes cisternasii*) in Spain (J. Bosch, unpublished data); the fire salamander (*S. salamandra*) in Spain and France (Bosch and Martinez-Solano 2006, M.C. Fisher, S.F. Walker, J. Bosch, unpublished data); the Tyrrhenian painted frog (*Discoglossus sardus*) in Italy (Bielby *et al.* 2009); the common toad (*Bufo bufo*) in Spain, France, and the UK (Bosch and Martinez-Solano 2006; Garner *et al.* 2009; J. Bosch, S. Walker, T. Garner, J. Bielby, A. Cunningham, M. Fisher, unpublished data); the natterjack (*Epidalea calamita*) in Spain (Martinez-Solano *et al.* 2003) and the UK (P. Minting, A. Cunningham, unpublished data); the palmate newt (*Lissotriton helveticus*) in France (J. Bosch, S. Walker, T. Garner, J. Bielby, M. Fisher, unpublished data); the ribbed newt (*Pleurodeles waltl*) in Spain (J. Bosch unpublished data); the edible frog (*Pelophylax* kl. *esculentus*) in Germany (Mutschmann *et al.* 2007). Mortality of captive Appenine yellow-bellied toads (*Bombina pachypus*) (Stagni *et al.* 2004) in Italy has also been attributed to lethal chytridiomycosis. Despite a broad distribution of the fungus in Europe, clinical outbreaks involving mass mortality have not been extensively published, which could reflect poor reporting or detection. Lethal chytridiomycosis in European hosts is often characterized by mass mortality (Bosch *et al.* 2001) but, for the majority of documented outbreaks, other causes of death cannot be ruled out due to a lack of comprehensive *post mortem* examination. Recently metamorphosed juveniles and adult amphibians seem predisposed to clinical disease but negative effects of infections during larval stages are known (Bosch *et al.* 2001; Bosch and Martinez-Solano 2006; Bielby *et al.* 2009; Garner *et al.* 2009; Walker *et al.* 2010; Luquet *et al.* 2012). No data exist for unhatched, developing embryos or gametes, but these are unlikely targets for infection due to their lack of keratinized tissues.

Given the poor knowledge of the incidence of lethal chytridiomycosis, it is unsurprising that relatively little is known about the transmission dynamics that underpin the spread of infection and emergence of disease. The global distribution of *B. dendrobatidis* illustrates that, provided potential amphibian hosts are present, infection occurs (Fisher *et al.* 2009a). This means that *B. dendrobatidis* is capable of transmitting and persisting under a range of biological and environmental conditions, including numerous habitat types throughout Europe (Bosch *et al.* 2001; Martinez-Solano *et al.* 2003; Simoncelli *et al.* 2005; Bosch and Martinez-Solano 2006; Adams *et al.* 2008; Bovero *et al.* 2008; Federici *et al.* 2008; Walker *et al.* 2008; Bielby *et al.* 2009; Garner *et al.* 2009; Pasmans *et al.* 2010; Tobler and Schmidt 2010; Walker *et al.* 2010; Ohst *et al.* 2011; Sztatecsny and Glasner 2011). Laboratory studies have identified *in vitro* optimal growth temperatures of 17–25°C. The occurrence of infected amphibian populations (Bosch *et al.* 2001; Martinez-Solano *et al.* 2003; Simoncelli *et al.* 2005; Bosch and Martinez-Solano 2006; Adams *et al.* 2008; Bovero *et al.* 2008; Federici *et al.* 2008; Walker *et al.* 2008; Bielby *et al.* 2009; Garner *et al.* 2009; Pasmans *et al.* 2010; Tobler and Schmidt 2010; Walker *et al.* 2010; Ohst *et al.* 2011; Sztatecsny and Glasner 2011), mass mortalities (Bosch *et al.* 2001; Bielby *et al.* 2009; Garner *et al.* 2009; Walker *et al.* 2010), and population declines (Bosch *et al.* 2001, 2007) within Europe at temperatures outside this range suggests that the *in vitro* optimal temperature range does not fully describe the conditions under which *B. dendrobatidis* can be transmitted. Recent analyses of field data (J. Bielby, F. Clare, M. Fisher, unpublished data), experimental studies (Rachowicz and Briggs 2007), and laboratory investigations (Woodhams *et al.* 2008) indicate that *B. dendrobatidis* can survive and infect susceptible hosts at temperatures as low as 4°C. While *B. dendrobatidis'* life-history trade-offs may be partly responsible for persistence and transmission dynamics at low temperatures (Woodhams *et al.* 2008), transmission under these conditions may also be influenced by changes in the host's behaviour and/or immunity (Ribas *et al.* 2009), changes in pathogen infection strategy (Fenton and Hudson 2002), or local adaptation of different strains or lineages of *Bd* (Fisher *et al.* 2009b; Farrer *et al.* 2011).

Although infection ultimately occurs as a result of invasion of keratinized tissues by motile zoospores, empirical evidence on the relative importance of direct host-host contact versus contact between infectious zoospores and potential hosts is lacking. In contrast, theoretical studies have highlighted the potential importance of modes of transmission other than via direct host-host contact. Epidemiological models have shown how the existence of a saprobic, free-living stage of *B. dendrobatidis* would greatly increase infection of hosts and the likelihood of population declines and extinction of European species (Mitchell *et al.* 2008). If *B. dendrobatidis* were able to persist for

Fig. 31.5 Examples of deformed European amphibians. A. Female *Bufo bufo* lacking the left hind foot (apody). B. *Bombina variegata* lacking the left hind limb (ectromelia). C. Male *Ichthyosaura alpestris* with incomplete dorsal crest. D. *Salamandra salamandra terrestris* with extreme truncation of the tail.

extended periods when availability of resources was reduced (for example when host density or availability of water is very low), the fitness of an individual parasite would be increased under certain circumstances. Although trade-offs in life history and persistence outside the host have been recorded in other parasites (Fenton and Hudson 2002), little direct evidence for a long-lived, saprobic stage in the fungus' life history exists (but see Di Rosa *et al.* 2007).

Empirical evidence for transmission functions that most accurately describe *B. dendrobatidis* dynamics are lacking, but some studies suggest that within a single species system, the rate at which new infections occur reaches asymptotes at higher densities, which is typical of non-density dependent transmission (Rachowicz and Briggs 2007; Briggs *et al.* 2010). If transmission is frequency-dependent and occurs at low host densities, *B. dendrobatidis* should prove to be a persistent parasite. The most commonly cited examples of frequency-dependent transmission are sexually transmitted diseases, in which spatiotemporal variation in clustering of susceptible and infected hosts dictates transmission rates rather than absolute population size (Anderson and May 1986). Several amphibian behaviours are associated with spatially and temporally proscribed aggregative behaviour. Tadpoles cluster while foraging, thereby avoiding predation and/or maintaining a growth-optimal body temperature in shallow water; also adult amphibians may gather in the thousands during the breeding season. All of these aggregations may serve to maintain contact rates and transmission of infection regardless of population size. In so doing, these behaviours of the host may aid *B. dendrobatidis* in avoiding density thresholds and thus increase the likelihood of persistence of the parasite and of decline of the host population.

Because *B. dendrobatidis* (like many parasites discussed in this chapter) is a generalist parasite infecting hundreds of species in complex amphibian assemblages, there is a need to be precise in the definition of "host population". In a multispecies or multistage amphibian assemblage, a single focal species or life-history stage may not function as a discrete host population. If interspecific transmission does occur the density threshold may not be reliant on any one single species, but on the density of certain amphibian species or age-classes within the community, as has been shown for ranaviruses (Brunner *et al.* 2004). There is little empirical data on interspecific transmission of *B. dendrobatidis*, but this is largely a result of lack of research effort, and the theoretical importance of the mechanism seems clear. Despite the lack of studies on interspecific transmission, studies on single-species systems have highlighted some general principles that will be important in future efforts to understand the dynamics of transmission and the potential role of *B. dendrobatidis* in population decline and extinction of host species. One such general finding is evidence of thresholds of infection and their role in mortality and population decline. In the Sierra Nevada, a combination of long-term field data and multi-state models outlined how likelihood of extirpation of populations of *Rana muscosa* may be explained solely by transmission dynamics. In high-density populations, reinfection of individuals was found to result in proliferation of infection intensity past threshold levels above which individuals are more likely to suffer mortality (Briggs *et al.* 2010; Vredenburg *et al.* 2010). However, these thresholds are generally thought to be exceeded only when proliferation occurs quickly and before any reduction in transmission occurs as a result of reduced rate of infectious contact, a scenario that is most likely to occur when host density is very high. Although threshold levels may vary according to a number of factors, including species of host (Stockwell *et al.* 2010), host's population genetics (Luquet *et al.* 2012), virulence of the *Bd* strain (Retallick and Miera 2007; Fisher *et al.* 2009b; Farrer *et al.* 2011), immune status of the host population (Ribas *et al.* 2009), and a number of environmental factors (Bosch *et al.* 2007; Walker *et al.* 2010), what little experimental work has been conducted seems to support the generality of the mechanism (Stockwell *et al.* 2010). Further attempts to clarify the existence of infection thresholds will be an interesting avenue for future research.

The source of *B. dendrobatidis* in Europe is not known, but lethal chytridiomycosis in the wild is predominantly associated with one globally distributed and pandemic lineage (Farrer *et al.* 2011). All known lineages of *B. dendrobatidis* are represented in Europe, and it appears the most virulent lineage has been introduced multiple times (Farrer *et al.* 2011). Research to determine the routes of introductions is underway (www.bd-maps.eu/), but it is likely that the introduction of non-native amphibians and the translocations of native species have played important roles in determining the current distribution of *B. dendrobatidis* across Europe (Garner *et al.* 2006; Fisher and Garner 2007). Support for this comes from the introduction of a second, less virulent, but still potentially lethal lineage onto the island of Mallorca via an amphibian reintroduction programme. Infection of reintroduced animals arose through switching of hosts in captivity (Walker *et al.* 2008; Farrer *et al.* 2011). Infected amphibians are documented in all arms of the amphibian trade (*e.g.* Walker *et al.* 2008; van der Sluijs *et al.* 2011) and no existing national or European Union regulations are in place to regulate *B. dendrobatidis* infection in the amphibian trade beyond the reporting requirements associated with membership in the OIE (Schloegel *et al.* 2010). Limiting the introduction of the fungus into naïve populations should be a goal of all those working with amphibians or in amphibian habitats, although quarantine of sensitive, disease-free amphibian habitats for the sake of amphibian health seems improbable, and impossible to maintain for the long term. Reducing forcing of infection (Briggs *et al.* 2010) is the more appropriate and feasible option and should be achievable through the adoption of sensible practices. These should include proper disinfection protocols after visiting or before going to amphibian habitats, translocating or reintroducing confirmed disease-free animals, and prohibiting the trade in infected animals, either as food, for research, or as ornamental or zoo animals. Enforcing these procedures will require the development of local risk assessments to justify protocols, education of stakeholders, and the development of the legal instruments and methods of enforcement to ensure compliance. All of these steps require the input of all stakeholders before development and implementation.

V. Oomycetes

Oomycetes are a group of fungal-like organisms in the Chromista that are ubiquitous in terrestrial, marine, and freshwater environments. Although they are better known as major threats to plant species (Hardham 2005; Brasier and Webber 2010), they have become a serious issue in animal populations, causing extirpations of populations and major ecological and economic damage (Edgerton *et al.* 2004; Andrew *et al.* 2008). Oomycetes are only very distantly related to fungi, but are traditionally studied by mycologists and share some important morphological and developmental traits with fungi (Beakes *et al.* 2011). Like chytrid fungi, Oomycetes have motile aquatic zoospores, but Oomycete zoospores are bi-flagellated and develop incredibly quickly, sometimes within a matter of minutes (Walker and van West 2007). Again similar to fungi, Oomycetes can form large filamentous colonies composed of hyphae, and studies have shown that these stages may secrete both digestive and host-manipulative proteins (van West *et al.* 2008). The vast majority of known Oomycetes are parasitic, and collectively they infect many lineages of eukaryotes (Kamoun 2003; Beakes *et al.* 2011). Among the major groups of Oomycetes some Peronosporomycetidae are known to be pathogens of vertebrates in terrestrial systems, but only the freshwater-inhabiting Saprolegniales are known as serious pathogens of amphibians (van West *et al.* 2008; Beakes *et al.* 2011). The high diversity of these saprolegnian pathogens has only recently been recognized, and tools for identification of species are crude (Hulvey *et al.* 2007; Petrisko *et al.* 2008). With the banning of malachite green, there are few anti-oomycotic agents that can be applied on a large scale to treat epizootics (van West *et al.* 2008).

A. Saprolegniosis

Infection with water moulds of the family Saprolegniaceae, termed saprolegniosis, is commonly associated with mortality of amphibian eggs and developing larvae (Kiesecker and Blaustein 1997, 1995; Robinson *et al.* 2003; Fernández-Benéitez *et al.* 2008). Secondary infections of adult amphibians can be fatal and mortality in recently metamorphosed froglets has been shown experimentally (Romansic *et al.* 2007). Infection at epithelial lesions (*e.g.* bite wounds) (Walls and Jaeger 1987) is accomplished either via free-swimming zoospores or thick-walled oospores. Advanced infection is typified by a cottony white or grey matt of hyphae protruding from the wound, the oral opening, or gill slits, or encapsulating the body. Hyphae can be 2–40 microns in diameter, have parallel walls, are aseptate and branch occasionally. Infection can result in necrotic, ulcerated and reddened skin, possibly cutaneous hemorrhagic (Taylor 2001) and may ultimately involve various tissues, including bone, muscle, and nervous tissue (Frye and Gillespie 1989). Hyphae may also be detected on the skin's surface without any obvious signs of previous wounding. Infection of egg masses and developing embryos appears to occur even without superficial damage to the egg capsule or the embryonic dermis (Blaustein *et al.* 1994).

One of the most dramatic epizootics in eggs and larvae involved the death of approximately 95% of an estimated 2.5 million eggs deposited at a single breeding site in a single breeding season (Blaustein *et al.* 1994). Disease was episodic but persistent, perhaps reflecting the environmental ubiquity of Saprolegniaceae. Disease progressed in a wave-like pattern across egg masses, suggesting a dynamic other than random infection from environmental spores; other studies have shown transmission among egg clutches through hyphal growth (Robinson *et al.*, 2003). Indeed, eggs laid in communal masses seem predisposed to infection (Kiesecker and Blaustein 1997), but susceptibility of eggs may decrease with advancing development (Robinson *et al.*, 2003).

European species affected by saprolegniosis of eggs and/or larvae include *Epidalea calamita* in the wild in Spain, the Netherlands, and the UK (Strijbosch 1979; Banks and Beebee 1988; Fernández-Benéitez *et al.* 2008); *R. temporaria* in the wild in the Netherlands and the UK (Strijbosch 1979; Beattie *et al.* 1991); *R. arvalis*, *P.* kl. *esculentus*, *B. bufo*, and *P. fuscus* in the Netherlands (Strijbosch 1979), *R. arvalis* from Sweden and *B. bufo* and *R. temporaria* from the UK have also been proven to be susceptible experimentally (Robinson *et al.* 2003; Sagvik *et al.*, 2008a,b). Fertilized eggs of a variety of amphibian species, including eggs of European amphibian species (Czeczuga *et al.* 1998), can harbour numerous Saprolegniaceae without any evidence of saprolegniosis (Petrisko *et al.* 2008). Evidence for an effect of host genetics on susceptibility argues that coevolution may buffer hosts against serious declines due to saprolegniosis (Sagvik *et al.*, 2008b). The environmental prevalence of Saprolegniaceae, their common occurrence on eggs without any visible pathology, and the ability of amphibian populations to experience large losses of eggs and embryos from saprolegniosis without clear impact on population dynamics (Biek *et al.* 2002), draw into question the inclusion of saprolegniosis in this list. Notwithstanding, it is clear that Saprolegniaceae can impose substantial mortality on early life-history stages; in the Netherlands complete reproductive failure of up to four amphibian species attributable to saprolegniosis was recorded over four years at multiple sites (Strijbosch 1979). Direct and indirect interactions between virulence and environmental factors can exacerbate disease (Strijbosch 1979; Banks and Beebee 1988; Kiesecker and Blaustein 1995; Ruthig 2008). This includes unpredictable interactions with environmental pollutants (Lefcort *et al.* 1997; Romansic *et al.* 2006). Awareness of this suite of pathogens of amphibians and further study of their potential impacts are still very much warranted.

Diagnosis can be achieved by examining wet-mounted scrapings of skin and subsequent culturing. Initial diagnosis in the field can be achieved using a hand lens (Green and Converse 2005). Molecular diagnostics are available for detection and limited differentiation of Saprolegniaceaen species, isolates, or clades (Petrisko *et al.* 2008). For additional information refer to Green (2001), Taylor (2001), and Berger *et al.* (2009).

VI. Mesomycetozoea

The exact phylogenetic placement of this group of microorganisms remains contentious (Ragan *et al.* 1996; Mendoza *et al.* 2002). Mesomycetozoeans, previously referred to as DRIPs (*Dermocystidium*, 'rosette agent', *Ichthyophonus*, and *Psorospermium*), have traits suggestive of both fungi and choanoflagellates, which have led authors to place them at the so-called animal-fungal boundary (Mendoza *et al.* 2002). Irrespective of their evolutionary placement, the majority of mesomycetozoeans are parasites of animals and members of both recognized orders (Dermocystida and Ichthyophonida) infect amphibians (Pérez 1907; Herman 1984; Mikaelian *et al.* 2000; Mendoza *et al.* 2002; Raffel *et al.* 2006, 2008). Infections with ichthyophonid-like organisms have only been reported for North American amphibians; here treatment is restricted to dermocystid-like organisms. See Herman (1984), Mikaelian *et al.* (2000), and Raffel *et al.* (2006) and references therein for further information regarding infections of amphibians with members of the ichthyophonid branch of the Class Mesomycetozoea.

A. *Amphibiocystidium* sp.

Dermocystid infections have been reported in European amphibians numerous times for over a century (Pérez 1907; Pérez 1913; Guyénot and Naville 1922; Remy 1931; Poisson 1937; Broz and

Prívora 1952) but only twice in the New World (Carini 1940; Raffel *et al.* 2008). Pascolini *et al.* (2003) reviewed the published literature for European cases and concluded that established taxonomy was inconsistent and based on inadequate criteria. They proposed a simplification of the taxonomy of European dermocystids infecting amphibians and erected a new genus, *Amphibiocystidium*. While such a classification is presumptuous based on a lack of evidence (as stated by the authors themselves), preliminary data on DNA sequences support, albeit weakly, the monophyly and derived status of European dermocystids infecting amphibians [but not that of all *Amphibiothecum* (*Dermosporidium*) sp. (see González-Hernández *et al.* 2010)]. In this chapter the convention of Pascolini *et al.* (2003) is adopted and refers to the dermocystid agents of infections of European amphibians as *Amphibiocystidium*.

The infection cycle has yet to be described, but the fact that *Amphibiocystidium* spp. form spores in the dermis of hosts is suggestive. Mendoza *et al.* (2002) described a putative life cycle of dermocystids in which *in vitro* endospores give rise to uniflagellate zoospores that infect hosts, encyst, grow, and, through cleavage, produce endospores. Additionally, they proposed that endospores can be released *in vivo* and the replication cycle can be completed without the parasite exiting the host. This depiction is challenged by a lack of detection and description of zoospores for European dermocystids of amphibians (Pascolini *et al.* 2003; González-Hernández *et al.* 2010; but see Poisson 1937). However, reports of *Amphibiocystidium* are of infections of hosts, which would not reveal environmentally developed zoospores. How entry into an uninfected host's dermis is accomplished is unknown, but infections by *Ichthyophonus* spp. appear to be facilitated by leeches (Raffel *et al.* 2006). Infection of metamorphosed amphibians manifests as vesicles, pre-ulcerative or ulcerative skin lesions, and nodular skin cysts of various shapes and sizes distributed across the body, limbs, tail, and head of the host (Pascolini *et al.* 2003; Raffel *et al.* 2008; González-Hernández *et al.* 2010). In one report, cysts appeared carbunculate as well as nodular (González-Hernández *et al.* 2010). Cysts can disseminate, and are documented as affecting internal organs such as the liver (Green and Sherman 2001; Raffel *et al.* 2008; A. Gray, T. Garner, K. Acevedo-Whitehouse unpublished data). Histological examination of cysts commonly reveals septate structures with chambers containing basophilic granulate spores, approximately 10 μm in diameter, clustered according to developmental stage (Broz and Prívora, 1952; Jay and Pohley 1981; Table 2 in Pascolini *et al.* 2003; González-Hernández *et al.* 2010).

Until recently, *Amphibiocystidium* infections were generally accepted as non-pathogenic (Guyénot and Naville, 1922; Pascolini *et al.* 2003). Some early authors reported morbidity and mortality (Moral 1913; Gambier 1924) but, without the benefit of clear diagnostics, the attribution of disease to an *Amphibiocystidium* parasite in these cases is uncertain. More recently, mortality of wild-collected and visibly infected North American eastern red-spotted newts (*Notophthalmus viridescens*) was observed in captivity post-collection (Raffel *et al.* 2008). Worrying data for Europe are also beginning to accumulate. Declines of pool frogs (*Pelophylax lessonae*) in Italy have been linked to *Amphibiocystidium* at one location (Pascolini *et al.* 2003) but no other recent evidence of European anuran mortality is clearly attributable to *Amphibiocystidium*. In contrast, data for European caudates supports the hypothesis that infections can be lethal. *I. alpestris* collected near Zürich, Switzerland, exhibited dermal lesions and experienced substantial mortality after several weeks in captivity. Lesions were diagnosed as *Amphibiocystidium* infections (Oevermann *et al.* 2005; Hoeck and Garner 2007). A similarly virulent outbreak in the same species was observed in animals collected from an invasive population in the UK (J. Sears, T. Garner, unpublished data). *L. helveticus* inhabiting an island in Scotland were found dead and dying in the spring of 2011, exhibiting extremely pronounced infections (Figs. 3, 4; L. Anderson, unpublished data). Post-mortem examinations are pending but, in a previous study of palmate newts at this location, lesions matching those seen in

2011 were caused by a genetically novel *Amphibiocystidium* (A. Gray, T. Garner, A. Meredith, K. Acevedo-Whitehouse, unpublished data). Both the severity of an individual infection and its prevalence in the population should relate to the pathogenicity of *Amphibiocystidium* (Pascolini *et al.* 2003: Raffel *et al.* 2008) and there is anecdotal evidence for recent increases in prevalence of infection in susceptible newts (González-Hernández *et al.* 2010).

Infections with *Amphibiocystidium* have been diagnosed in *P.* kl. *esculentus* and *P. lessonae* in Italy and France (Remy 1931; Pascolini *et al.* 2003); *A. obstetricans* in France and Switzerland (Pérez 1913; Guyénot and Naville 1922); *R. temporaria* in Switzerland and the then Czechoslovakia (Guyénot and Naville 1922; Broz and Prívora 1952); *R. arvalis* in Switzerland (T. Garner personal observation); *L. helveticus* in France and the UK (Poisson 1937; Gonzalez-Hernandez *et al.* 2010; F. Smith, T. Garner, and others, personal observations; A. Gray, T. Garner, A. Meredith, K. Acevedo-Whitehouse, unpublished data); marbled newts (*Triturus marmoratus*) in France (Pérez 1907); great crested newts (*T. cristatus*) in France and Belgium (Pérez 1913; A. Martel, F. Pasmans, unpublished data); *L. vulgaris* and *I. alpestris* in Belgium (A. Martel, F. Pasmans, unpublished data); *I. alpestris* in Switzerland and the UK (Oevermann *et al.* 2005; Hoeck and Garner 2007; T. Garner and J. Sears unpublished data). Clearly, many, if not the majority, of these cases have not resulted in high mortality rates, but determining if lethal infections are occurring and if infection is on the increase is hampered by the lack of structured investigation and the fact that many affected newts are in the breeding phase. During this period, European newt species of the genera *Triturus*, *Lissotriton* and *Icthyosaura* (including the two lethally affected species listed above) are aquatic and cryptic, making the detection of mortality extremely difficult. Interestingly, the aquatic nature of courting newts may be linked to pathogenicity. Infected alpine newts provided with a choice of aquatic and terrestrial habitats sometimes moved onto land, where in some cases skin lesions would disappear over relatively short periods of time (P. Hoeck, unpublished data). Moving to a terrestrial environment may represent a non-immunological host response (Murphy *et al.* 2011; Parker *et al.* 2011). This may not be the only response available to amphibians confronted with mesomycetozoean exposures and infections. Development of an acquired immune response against *Amphibiocystidium* infection has been suggested by Raffel *et al.* (2009).

Diagnosis of mesomycetozoean infections includes the identification of gross skin lesions as described above. Lesions may be widely dispersed across the body, but they often cluster at junctions of limbs, on digits, along limbs and tails, and at cloacal openings (Fig. 31.4). Fresh and formalin-fixed tissues can be used for diagnosis via histology (HE or Giemsa staining). Histology should be accompanied with electron microscopic examinations of lesions, along with molecular diagnostics (see González-Hernández *et al.* 2010). Currently no quantitative PCR exists for mesomycetozoean parasites and all attempts to culture amphibian mesomycetozoeans have failed.

VII. Protozoa

Protozoa are single-celled, often motile eukaryotes, many of which are responsible for a wide variety of infectious diseases of warm-blooded vertebrates. They are also ubiquitous parasites of larval and metamorphosed amphibians. Diseases caused by several types of protozoa have been reported in European amphibians. Geographically widespread infection with the coccidian parasite, *Goussia* spp., in larvae of several European anurans (*P.* kl. *esculentus*, *R. dalmatina*, *R. temporaria*, *B. bufo*, green toads *Pseudepidalea viridi*) has been reported recently (Jirků *et al.* 2009a,b; see references therein for older case studies). Infected larvae that were euthanized and dissected had obvious signs of disease (coccidiosis), including substantial necrosis of the intestinal epithelium, yet infected larvae which shed quantities of oocysts comparable to those shed by dissected larvae survived and completed metamorphosis. The authors concluded that, although occurring at high prevalence and capable of causing serious lesions, the coccidian species they described likely do not pose a significant threat to European amphibian populations.

In mammals, apicomplexan protozoa often are species-specific or genus-specific parasites (*e.g.* Pedersen *et al.* 2005). Amphibian apicomplexans infecting European hosts seem to be at least genus specific (Jirků *et al.* 2009a; 2009b). A non-native and potentially invasive apicomplexan (*Cryptosporidium fragile*) isolated from newly imported Malaysian black-spined toads, *Duttaphrynus melanostictus*, could not be experimentally transmitted to European anurans (Jirků *et al.* 2008). Taken together, these findings suggest that most apicomplexan protozoan dynamics observed in European amphibians are not of recent origin and the incursion of novel apicomplexans is unlikely. Further work is required to test these hypotheses and, as for the control of other pathogens, there remains a requirement for strict quarantine measures and import regulation for amphibians so as to prevent introductions of protozoan parasites (Jirků *et al.* 2008).

Myxozoan parasites are common sporozoan parasites of amphibians that affect both European caudates and anurans (Eiras 2005). Again, recent data show that while infection may be common and may cause internal lesions, infected animals exhibit no obvious outward signs of disease (Jirků *et al.* 2007). Overall, no examples of substantial mortality of wild European amphibians due to protozoan infections have been reported, but this may merely be because of lack of study. The fact so many protozoans cause notable disease in other vertebrate classes suggests a need for more comprehensive investigations of protozoan parasites of amphibians.

VIII. Metazoa

Parasitic platyhelminths (flatworms) are bilaterally symmetrical, acoelomate, unsegmented eukaryotes that infect their hosts internally. Flatworms are predominantly the Cestoda (tapeworms), the Trematoda (flukes), and the Monogenea. Along with the Phyla Acanthocephala and Nematoda, they are common, globally distributed parasites of amphibians (Goater 1992; Barton 1999; Kuzmin *et al.* 2003; Yildirimhan *et al.* 2005; Prudhoe and Bray 1982; McKenzie 2007). All three phyla include species that are transmitted directly or via vectors and, depending on the parasite's life history, amphibians may act either as intermediate or definitive hosts (Prudhoe and Bray 1982; Moravic and Škoríková 1998; Bolek *et al.* 2010). European amphibians are commonly infected with platyhelminth, acanthocephalic, and nematode parasites and there is published evidence of indirect and direct effects of infection by a lung nematode (*Rhabdias bufonis*) on components of fitness, including survival, in *B. bufo* (Goater *et al.* 1993; Goater and Ward 1994). Accumulating evidence for platyhelminth infection as a significant disease and conservation issue in amphibians comes from studies of digenean trematodes in North America, focussed on three genera: *Clinostomum*, *Echinostoma* and *Ribeiroia* (Johnson *et al.* 1999, 2003, 2006, 2007; Schotthoefer *et al.* 2003a; Johnson and Chase 2004; Miller *et al.* 2004; Holland *et al.* 2007; Nieto *et al.* 2007; Rohr *et al.* 2008a,b; McAllister *et al.* 2010). Although published information on costs to hosts predominantly implicate *Echinostoma* and *Ribeiroia* as causes of morbidity, mortality, or declines of host species (Johnson *et al.* 1999; Holland *et al.* 2007; Rohr *et al.* 2008a,b), infection dynamics of *Clinostomum*, as discussed by Gray *et al.* (2007), mimic those of *Ribeiroia* and there is evidence that *Clinostomum* infection can cause disease. For the purposes of this review, reference is made to trematodes and no distinction is made amongst these three genera.

Trematode metacercariae are transmitted from intermediate snail hosts to amphibian hosts, and via these to the definitive avian or mammalian hosts (Prudhoe and Bray 1982). Metacercariae encysting in adult amphibians are not considered to be seriously costly to the host; however, encystment in larval amphibians can have serious consequences. Depending on where encystment occurs and the developmental stage of the larvae, parasitism may cause a wide range of developmental abnormalities, including anophthalmy, ectromelia, apody, and ecrodactyly, polymelia, polypody, polydactyly, and cutaneous fusion, as well as significantly reducing the probability of larval survival (Johnson *et al.* 1999; Schotthoefer *et al.* 2003a,b). The dynamics of infection are strongly affected by the use of fertilizers, which cause eutrophication of ponds, population increases of snail hosts, and subsequent increases in rates of transmission to susceptible larvae (Johnson and Chase 2004; Johnson *et al.* 2007; Rohr *et al.* 2008a,b). This has been linked to local declines of amphibian populations (Rohr *et al.* 2008b).

Given the broad distribution of fertilizer-intensive agricultural practices in Europe, it is reasonable to assume that an amphibian-host/trematode-parasite system similar to that described in North America may occur in Europe. Links between developmental abnormalities and agricultural pollutants in the UK are starting to be made (Orton and Routledge 2011) but the additional link to platyhelminth infection has not. Amphibians exhibiting developmental abnormalities are certainly found in Europe (Fig. 31.5), but is there any evidence that deformities are of concern for amphibian conservation in Europe? Piha *et al.* (2006) detected a low frequency of abnormalities in recently metamorphosed *R. temporaria* but no evidence that the frequency was related to probability of exposure to agrochemicals. This is in contrast to the report by Puky (2006), in which extremely high local frequencies of abnormalities were related to tadpole density, higher water temperature, concentrations of pesticides, and bacterial infections. These studies did not report any data on presence or prevalence or metacercariae. Again, further investigations are recommended and caution is urged against ignoring the possibility that this suite of parasites may be affecting population health and demographics of amphibians in Europe.

Diagnosis of infections requires a combination of microscopy, histopathology, and, when limb abnormalities are detected, careful radiography (Johnson *et al.* 1999; Schotthoefer *et al.* 2003a,b; Lannoo 2008). Supernumerary limbs, feet, and digits, as well as other skeletal and morphological abnormalities, may occur for a variety of reasons other than parasitism (Sessions and Ruth 1990; Blaustein and Johnson 2003; Lannoo 2008; Johnson and Bowerman 2010) and missing extremities and other forms of deformities are often the result of damage inflicted by predators (Ballengée and Sessions 2009).

IX. Concluding remarks

The diversity of parasites reported in or on European amphibians is as broad as seen in amphibians in any other part of the world. Nevertheless, there is a lack of quantitative ecological and epidemiological research on parasites infecting Europe's amphibians: the majority of what has been published is biased towards two notable pathogens (*Ranavirus* spp. and *B. dendrobatidis*). It remains unclear what conditions may allow a parasite to become lethal to a European anuran or caudate and even in cases when costs to hosts appear straightforward, the lack of post-mortem examinations leave questions as to the relative contribution a parasite makes to hosts' morbidity and mortality. What is clear is that at least some parasites of amphibians occurring in Europe have the capacity to cause persistent population declines and may be threats to the persistence of amphibian species (Berger *et al.* 1998; Daszak *et al.* 2003; Teacher *et al.* 2010), as is true for parasites of other vertebrate hosts (Atkinson *et al.* 1995; Frick *et al.* 2010; Robinson *et al.* 2010).

The rich history of European herpetology may provide a unique opportunity for developing a better understanding of parasite dynamics and how they may link to amphibian population dynamics. Some of the oldest and relatively continuous datasets on amphibian ecology are on European species (Houlahan *et al.* 2000) and some of the earliest descriptions of amphibian parasites are of those infecting European hosts (Leeuwenhoek 1706; Frölich 1789; Vulpian 1859; Pérez 1907, 1913). European archives house substantial collections of preserved amphibians and as the other chapters in this issue show, national monitoring schemes for amphibians are common across the continent. The precedence for developing European-level projects investigating the implications for conservation of amphibian parasites is already set (http://www.bd-maps.eu/). Now that two amphibian infectious diseases require reporting at the national level (Schloegel *et al.* 2010), federal agencies and laboratories should already be engaged in diagnostic *post-mortem* examinations of amphibian mortality events. The pieces are in place; all that is needed is for the various stakeholders to develop the collaborations. All European amphibian monitoring schemes should include an infectious disease component and, whenever possible, so should ecological studies of amphibian populations. Such studies inevitably yield interesting and often scientifically important findings, but given the current diseases affecting amphibians at a global scale, they may well also yield information crucial for the conservation of European amphibians.

X. References

Adams, M. J., Galvan, S., Scalera, R., Grieco, C. and Sindaco, R., 2008. *Batrachochytrium dendrobatidis* in amphibian populations in Italy. *Herpetological Review* **39**: 324–326.

Aguirre, A. A. and Lutz, P. L., 2004. Marine turtles as sentinels of ecosystem health: is fibropapillomatosis an indictor? *Ecohealth* **1**: 275–283.

Ahne, W., Bearzotti, M., Bremont, M. and Essbauer, S., 1998. Comparison of European systemic piscine and amphibian iridoviruses with Epizootic Haematopoietic Necrosis virus and Frog Virus 3. *Journal of Veterinary Medicine Series B* **45**: 373–383.

Anderson, P. K., Cunningham, A. A., Patel, N. G., Morales, F. J., Epstein, P. R. and Daszak, P., 2004. Emerging infectious diseases of plants: pathogen pollution, climate change and agrotechnology drivers. *Trends in Ecology and Evolution* **19**: 435–544.

Anderson, R. M. and May, R. M., 1986. The invasion, persistence and spread of infectious diseases within animal and plant communities. *Philosophical Transactions of the Royal Society of London, Series B* **314**: 533–570.

Andre, S. E., Parker, J. and Briggs, C. J., 2008. Effect of temperature on host response to *Batrachochytrium dendrobatidis* infection in the mountain yellow-legged frog (*Rana muscosa*). *Journal of Wildlife Diseases* **44**: 716–720.

Andrew, T. G., Huchzermeyer, K. D. A., Mbeha, B. C. and Nengu, S. M., 2008. Epizootic ulcerative syndrome affecting fish in the Zambezi river system in southern Africa. *Veterinary Record* **163**: 629–632.

Araújo, M. B., Thuiller, W. and Pearson, R. G., 2006. Climate warming and the decline of amphibians and reptiles in Europe. *Journal of Biogeography* **33**: 1712–1728.

Ariel, E., Kielgast, J., Svart, H. E., Larsen, K., Tapiovaara, H., Jensen, B. B. and Holopainen, R., 2009. Ranavirus in wild edible frogs *Pelophylax* kl. *esculentus* in Denmark. *Diseases of Aquatic Organisms* **85**: 7–14.

Atkinson, C. T., Woods, K. L., Dusek, R. J., Sileo, L. S. and Iko, W. M., 1995. Wildlife disease and conservation in Hawaii: pathogenicity of avian malaria (*Plasmodium relictum*) in experimentally infected liwi (*Vestiaria coccinea*). *Parasitology* **111**: S59–S69

Ballengée, B. and Sessions, S. K., 2009. Explanation for missing limbs in deformed amphibians. *Journal of Experimental Zoology* **312B**: 1–10.

Balseiro, A., Dalton, K. P., del Cerro, A., Márquez, I., Cunningham, A. A., Parra, F., Prieto, J. M. and Casais, R., 2009. Pathology, isolation and molecular characterization of a ranavirus from the common midwife toad (*Alytes obstetricans*) on the Iberian Peninsula. *Diseases of Aquatic Organisms* **84**: 95–104

Balseiro, A., Dalton, K. P., del Cerro, A., Márquez, I., Parra, F., Prieto, J. M. and Casais, R., 2010. Outbreak of common midwife toad virus in alpine newts (*Mesotriton alpestris cyreni*) and common midwife toads (*Alytes obstetricans*) in Northern Spain: a comparative pathological study of an emerging ranavirus. *Veterinary Journal* **186**: 256–258.

Banks, B. and Beebee, T. J. C., 1988. Reproductive success of natterjack toads *Bufo calamita* in two contrasting habitats. *Journal of Animal Ecology* **57**: 475–492.

Barton, D. P., 1999. Ecology of helminth communities in tropical Australian amphibians. *International Journal of Parasitology* **29**: 921–926.

Beakes, G. W., Glockling, S. L., and Sekimoto, S., 2011. The evolutionary phylogeny of the oomycete "fungi". *Protoplasma* **249**: 3–19.

Beattie, R. C., Aston, R. J. and Milner, A. G. P., 1991. A field study of fertilization and development in the common frog *Rana temporaria* with particular reference to acidity and temperature. *Journal of Applied Ecology* **28**: 346–357.

Becker, C. G. and Zamudio, K. R., 2012. Tropical amphibian populations experience higher disease risk in natural habitats. *Proceedings of the National Academy of Sciences of the USA*, **108**: 9893–9898.

Bennati, R., Bonetti, M., Lavazza, A. and Gemetti, D., 1994. Skin lesions associated with herpesvirus-like particles in frogs (*Rana dalmatina*). *Veterinary Record* **135**: 625–626.

Berger, L., Speare, R., Daszak, P., Green, D. E., Cunningham, A. A., Goggin, C. L., Slocombe, R., Ragan, M. A., Hyatt, A. D., McDonald, K. R., Hines, H. B., Lips, K. R., Marantelli, G. and Parkes, H., 1998. Chytridiomycosis causes amphibian mortality associated with population declines in the rain forests of Australia and Central America. *Proceedings of the National Academy of Sciences of the USA* **95**: 9031–9036.

Berger, L., Marantelli, G., Skerratt, L. L. and Speare, R., 2005. Virulence of the amphibian chytrid fungus *Batrachochytrium dendrobatidis* varies with the strain. *Diseases of Aquatic Organisms* **68**: 47–50.

Berger, L., Longcore, J. E., Speare, R., Hyatt, A. and Skerratt, L. L., 2009. Fungal diseases of amphibians. Pp. 2986–3052 in *Amphibian Biology, Volume 8 - Amphibian Decline: Diseases, Parasites, Maladies and Pollution*, ed by H. Heatwole and J. W. Wilkinson. Surrey Beatty and Sons, Baulkham Hills.

Biek, R., Funk, W. C., Maxell, B. A. and Mills, L. S., 2002. What is missing in amphibian decline research: insights from ecological sensitivity analysis. *Conservation Biology* **16**: 728–734.

Bielby, J., Bovero, S., Sotgiu, G., Tessa, G., Favelli, M., Angelini, C., Doglio, S., Clare, F., Gazzaniga, E., Lapietra, F. and Garner, T. W. J., 2009. Fatal chytridiomycosis in the Tyrrhenian painted frog. *Ecohealth* **6**: 27–32.

Bielby, J., Bosch, J., Churcher, T. S., Clare, F. C., Garner, T. W. J., Schmeller, D. S., Walker, S. F. and Fisher, M. C. Seasonal variation in transmission dynamics explains heterogeneity in host competence for a host generalist pathogen. *Proceedings of the Royal Society of London, Series B.* [in review]

Blackwell, M., 2011. The Fungi: 1, 2, 3 ... 5.1 Million Species? *American Journal of Botany* **98**: 426–438.

Blaustein, A. R. and Johnson, P. T. J., 2003. The complexity of deformed amphibians. *Frontiers in Ecology and the Environment* **2**: 87–94.

Blaustein, A. R., Hokit, G. R., O'Hara, R. K. and Holt, R. A., 1994. Pathogenic fungus contributes to amphibian losses in the Pacific Northwest. *Biological Conservation* **67**: 251–254.

Blaustein, A. R., Han, B. A., Relyea, R. A., Johnson, P. T. J., Buck, J. C., Gervasi, S. S. and Kats, L. B., 2011. The complexity of amphibian population declines: understanding the role of cofactors in driving amphibian losses. *Annals of the New York Academy of Sciences* **1223**: 108–119.

Blumer, C., Zimmermand, D.R., Weilenmann, R., Vaughan, L. and Pospischil, A., 2007. *Chlamydiae* in free-ranging and captive frogs in Switzerland. *Veterinary Pathology* **44**: 144–150.

Bolek, M. G., Tracy, H. R. and Janovy Jr., J., 2010. The role of damselflies (Odonata: Zygoptera) as paratenic hosts in the transmission of *Halipegus eccentricus* (Digenea: Hemiuridae) to anurans. *Journal of Parasitology* **96**: 724–735.

Bosch, J., and Martinez-Solano, I., 2006. Chytrid fungus infection related to unusual mortalities of *Salamandra salamandra* and *Bufo bufo* in the Penalara Natural Park, Spain. *Oryx* **40**: 84–89.

Bosch, J., Martinez-Solano, I. and Garcia-Paris, M., 2001. Evidence of a chytrid fungus infection involved in the decline of the common midwife toad (*Alytes obstetricans*) in protected areas of central Spain. *Biological Conservation* **97**: 331–337.

Bosch, J., Rincón, P. A., Boyero, L. and Martínez-Solano, I., 2006. Effects of introduced salmonids on a montane population of the Iberian frog. *Conservation Biology* **20**: 180–189

Bosch, J., Carrascal, L. M., Duran, L., Walker, S. and Fisher, M. C., 2007. Climate change and outbreaks of amphibian chytridiomycosis in a montane area of Central Spain: is there a link? *Proceedings of the Royal Society, Series B* **274**: 253–260.

Bovero, S., Sotgiu, G., Angelini, C., Doglio, S., Gazzaniga, E., Cunningham, A. A. and Garner, T. W. J., 2008. Detection of chytridiomycosis caused by *Batrachochytrium dendrobatidis* in the endangered Sardinian newt

(*Euproctus platycephalus*) in southern Sardinia, Italy. *Journal of Wildlife Diseases* **44**: 712–715.

Boyle, D. G., Boyle, D. B., Olsen, V., Morgan, J. A. T. and Hyatt, A. D., 2004. Rapid quantitative detection of chytridiomycosis (*Batrachochytrium dendrobatidis*) in amphibian samples using real-time Taqman PCR assay. *Diseases of Aquatic Organisms* **60**: 133–139.

Bradford, D. F., 1991. Mass mortality and extinction in a high elevation population of *Rana muscosa*. *Journal of Herpetology* **25**: 369–377.

Brasier, C., and Webber, J., 2010. Plant pathology sudden larch death. *Nature* **466**: 824–825.

Braunwald, J., Nonnenmacher, H. and Tripier-Darcy, F., 1985. Ultrastructural and biochemical study of frog virus 3 uptake by BHK-21 cells. *Journal of General Virology* **66**: 283–293.

Briggs, C. J., Knapp, R. A. and Vredenburg, V. A., 2010. Enzootic and epizootic dynamics of the chytrid fungal pathogen of amphibians. *Proceedings of the National Academy of Sciences of the USA* **107**: 9695–9700.

Broz, O. and Prívora, M., 1952. Two skin parasites of *Rana temporaria*: *Dermocystidium ranae* Guyénot & Naville and *Dermosporidium granulosom* n. sp. *Parasitology* **42**: 65–69.

Brucker, R. M., Baylor, C. M., Walters, R. L., Lauer, A., Harris, R. N. and Minbiole, K. P. C., 2008a. The identification of 2,4-diacetylphloroglucinol as an antifungal metabolite produced by cutaneous bacteria of the salamander *Plethodon cinereus*. *Journal of Chemical Ecology* **34**: 39–43.

Brucker, R. M., Harris, R. N., Schwantes, C. R., Gallaher, T. N., Flaherty, D. C., Lam, B. A. and Minbiole, K. P. C., 2008b. Amphibian chemical defense: antifungal metabolites of the microsymbiont *Janthinobacterium lividum* on the salamander *Plethodon cinereus*. *Journal of Chemical Ecology* **34**: 1422–1429.

Brunner, J. L., Schock, D. M., Davidson, E. W. and Collins, J. P., 2004. Intraspecific reservoirs: complex life history and the persistence of a lethal ranavirus. *Ecology* **85**: 560–566.

Brunner, J. L., Schock, D. M. and Collins, J. P., 2007. Transmission dynamics of the amphibian ranavirus *Ambystoma tigrinum* virus. *Diseases of Aquatic Organisms* **77**: 87–95.

Carey, C., 1993. Hypothesis concerning the causes of the disappearance of boreal toads from the mountains of Colorado. *Conservation Biology* **7**: 355–362.

Carlson, D. L., Rollins-Smith, L. A. and McKinnell, R. G., 1994. The Lucké herpesvirus genome: its presence in neoplastic and normal kidney tissue. *Journal of Comparative Pathology* **110**: 349–355.

Carini, A., 1940. Sobre um parasito semelhante ao "*Rhinosporidium*" encontrado em quistos da pelede uma "*Hyla*". *Arquivos do Institutos Biologico* **11**: 93–98.

Carver, S., Bell, B. D. and Waldman, B., 2010. Does chytridiomycosis disrupt amphibian skin function? *Copeia* **2010**: 487–495.

Chai, N., Deforges, L., Sougakoff, A., Truffot-Pernot, C., De Luze A., Demeneix, B., Clement, M. and Bomsel, M.C., 2006. *Mycobacterium szulgai* infection in a captive population of African clawed frogs (*Xenopus tropicalis*). *Journal of Zoo and Wildlife Medicine* **37**: 55–58.

Cheng, T. L., Rovito, S. M., Wake, D. B. and Vredenburg, V. T., 2011. Coincident mass extirpation of neotropical amphibians with the emergence of the infectious fungal pathogen *Batrachochytrium dendrobatidis*. *Proceedings of the National Academy of Sciences of the USA* **108**: 9502–9507.

Chinchar, V. G. and Dholakia, J. N., 1989. Frog virus 3-induced translational shut-off: activation of an eIF-2 kinase in virus-infected cells. *Virus Research* **14**: 207–224.

Chinchar, V. G. and Granoff, A., 1984. Isolation and characterization of a frog virus 3 variant resistant to phosphonoacetate: genetic evidence for a virus-specific DNA polymerase. *Virology* **138**: 357–361.

Chinchar, V. G. and Granoff, A., 1986. Temperature-sensitive mutants of frog virus 3: biochemical and genetic characterization. *Journal of Virology* **58**: 192–202.

Chinchar, V. G., Hyatt, A., Miyazaki, T. and Williams, T., 2009. Family *Iriodviridae*: Poor

Viral Relations No Longer. Pp. 123–170 in "Lesser Kown Large dsDNA Viruses", ed by J. L. Van Etten. Current Topics in Microbiology and Immunology **328**. Springer-Verlag, Berlin Heidelberg.

Crawford, A. J., Lips, K. R. and Bermingham, E., 2010. Epidemic disease decimates amphibian abundance, species diversity, and evolutionary history in the highlands of central Panama. *Proceedings of the National Academy of Sciences of the USA* **107**: 13777–13782

Cunningham, A. A., Langton, T. E. S., Bennett, P. M., Lewin, J. F., Drury, S. E. N., Gough, R. E. and MacGregor, S. K., 1996. Pathological and microbiological findings from incidents of unusual mortality of the common frog (*Rana temporaria*). *Philosophical Transactions of the Royal Society, Series B* **351**: 1539–1557.

Cunningham, A. A., Hyatt, A. D., Russell, P. and Bennett, P. M., 2007a. Experimental transmission of a ranavirus disease of common toads (*Bufo bufo*) to common frogs (*Rana temporaria*). *Epidemiology and Infection* **135**: 1213–1216.

Cunningham, A. A., Hyatt, A. D., Russell, P. and Bennett, P. M., 2007b. Emerging epidemic diseases of frogs in Britain are dependent on the source of ranavirus agent and the route of exposure. *Epidemiology and Infection* **135**: 1200–1212.

Cunningham, A. A., Tems, C. A. and Russell, P. H., 2008. Immunohistochemical demonstration of ranavirus antigen in the tissues of infected frogs (*Rana temporaria*) with systemic haemorrhagic or cutaneous ulcerative disease. *Journal of Comparative Pathology* **138**: 3–11.

Czeczuga, B., Muszyńska, E. and Krzemińska, A., 1998. Aquatic fungi growing on the spawn of certain amphibians. *Amphibia-Reptilia* **19**: 239–251.

Daszak, P., Cunningham, A. A. and Hyatt, A. D., 2003. Infectious disease and amphibian population declines. *Diversity and Distributions* **9**: 141–150.

Davison, A. J., Sauerbier, W., Dolan, A., Addison, C. and McKinnell, R. G., 1999. Genomic studies of the Lucké tumor herpesvirus (RaHV-1). *Journal of Cancer Research and Clinical Oncology* **125**: 232–238.

De Dastro, F. and Bolker, B., 2005. Mechanisms of disease-induced extinction. Ecology Letters **8**: 117–126.

Denoël, M., Dzukic, G. and Kalezic, M. L., 2005. Effects of widespread fish introductions on paedomorphic newts in Europe. *Conservation Biology* **19**: 162–170.

Densmore, C. L. and Green, D. E., 2007. Diseases of amphibians. *ILAR Journal* **48**: 235–254.

Di Rosa, I., Simoncelli, F., Fagotti, A. and Pascolini, R., 2007. The proximate cause of frog declines? *Nature* **447**: E4–E5.

Duffus, A. L. J., 2010. Chytrid blinders: what other disease risks to amphibians are we missing? *Ecohealth* **6**: 335–339.

Duffus, A. L. J. and Cunningham, A. A., 2010. Major disease threats to European amphibians. *Herpetological Journal* **20**: 117–127.

Duffus, A. L. J., Pauli, B. D., Wozney, K., Brunetti, C. R. and Berrill, M., 2008. Frog virus 3- like infections in aquatic amphibian communities. *Journal of Wildlife Diseases* **44**: 109–120.

Edgerton, B. F., Henttonen, P., Jussila, J., Mannonen, A., Paasonen, P., Taugbol, T., Edsman, L., and Souty-Grosset, C., 2004. Understanding the causes of disease in European freshwater crayfish. *Conservervation Biology* **18**: 1466–1474.

Eiras, J. C., 2005. An overview on the myxosporean parasites in amphibians and reptiles. *Acta Parasitologica* **50**: 267–275.

Emerson, H. and Norris, C., 1905. "Red-leg" - an infectious disease of frogs. *Journal of Experimental Medicine* **7**: 32–58.

Essbauer, S., Bremont, M. and Ahne, W., 2001. Comparison of the eIF-2 alpha homologous proteins of seven ranaviruses (*Iridoviridae*). *Virus Genes* **23**: 347–359.

Farrer, R. A., Weinert, L. A., Bielby, J., Garner, T. W. J., Balloux, F., Clare, F., Bosch, J., Cunningham, A. A., Weldon, C., du Preez, L. H., Anderson, L., Kosakovsky Pond, S. L., Shahar-Golan, R., Henk, D. A. and Fisher, M. C., 2011. Multiple emergences of amphibian chytridiomycosis include a globalised hypervirulent recombinant lineage. *Proceedings of the National Academy of Sciences of the USA* **108**: 18732–18736.

Federici, S., Clemenzi, S., Favelli, M., Tessa, G., Andreone, F., Casiraghi, M. and Crottini, A., 2008. Identification of the pathogen *Batrachochytrium dendrobatidis* in amphibian populations of a plain area in the northwest of Italy. *Herpetology Notes* **1**: 33–37.

Fenton, A. and Hudson, P. J., 2002. Optimal infection strategies: should macroparasites hedge their bets? *Oikos* **96**: 92–101.

Fernández-Benéitez, M. J., Ortiz-Santaliestra, M. E., Lizana, M. and Diéguez-Uribeono, J., 2008. *Saprolegnia diclina*: another species responsible for the emergent disease 'Saprolegnia infections' in amphibians. *FEMS Microbiology Letters* **279**: 23–29

Ficetola, G. F. and De Bernardi, F., 2004. Amphibians in a human-dominated landscape: the community structure is related to habitat features and isolation. *Biological Conservation* **119**: 219–230.

Ficetola, G. F., Padoa-Schioppa, E. and De Bernardi, F., 2009. Influence of landscape elements in riparian buffers on the conservation of semiaquatic amphibians. *Conservation Biology* **23**: 114–123.

Ficetola, G. F., Maiorano, L., Falcucci, A., Dendoncker, N., Boitani, L., Padoa-Schioppa, E., Miaud, C. and Thuiller, W., 2010. Knowing the past to predict the future: land-use change and the distribution of invasive bullfrogs. *Global Change Biology* **16**: 528–537.

Fijan, N., Matasin, Z., Petrinec, Z., Valpotic, I. and Zillenberg, L. O., 1991. Isolation of an iridovirus-like agent from the green frog (*Rana esculenta* L.). *Veterinarski Arhiv* **61**: 151–158.

Fisher, M. C. and Garner, T. W. J., 2007. The relationship between the introduction of *Batrachochytrium dendrobatidis*, the international trade in amphibians and introduced amphibian species. *Fungal Biology Reviews* **21**: 2–9.

Fisher, M. C., Garner, T. W. J. and Walker, S. F., 2009a. The global emergence of *Batrachochytrium dendrobatidis* in space, time and host. *Annual Review of Microbiology* **63**: 291–310.

Fisher, M. C., Bosch, J., Yin, Z., Stead, D. A., Walker, J., Selway, L., Brown, A. J. P., Walker, L. A., Gow, N. A. R., Stajich, J. E. and Garner, T. W. J., 2009b. Proteomic and phenotypic profiling of the amphibian pathogen *Batrachochytrium dendrobatidis* shows that genotype is linked to virulence. *Molecular Ecology* **18**: 415–429.

Forson, D. D. and Storfer, A., 2006. Atrazine increases ranavirus susceptibility in the tiger salamander, *Ambystoma tigrinum*. *Ecological Applications* **16**: 2325–2332.

Frank, W., 1976. Mycotic infections in amphibians and reptiles. Pp. 73–88 in *Proceedings of the 3rd International Wildlife Disease Conference*, ed by L. A. Page. Plenum Press, New York.

Frank, W., Roester, U. and Scholer, H. J., 1974. Sphaerule formation by a *Mucor* species in the internal organs of Amphibia. Pp. 161–192 in *A Century of Mycology*, ed by B. Sutton. Cambridge University Press, Cambridge.

Frick, W. F., Pollock, J. F., Hicks, A. C., Langwig, K. E., Reynolds, D. S., Turner, G. G., Butchkoski, C. M. and Kunz, T. H., 2010. An emerging disease causes regional population collapse of a common North American bat species. *Science* **329**: 679–682.

Frölich, J. A., 1789. Beschreibungen einiger neuer Eingeweidewürmer. *Der Naturforscher* **Part 24**: 101–162

Frohlich-Nowoisky, J., D. Pickersgill, A., Despres, V. R. and Poschl, U., 2009. High diversity of fungi in air particulate matter. *Proceedings of the National Academy of Sciences of the USA* **106**: 12814–12819.

Froufe, E., Arntzen, J. W. and Loureiro, A., 1999. Dead newts in Peneda-Gerês. *Froglog* **33**:1.

Frye, F. L. and Gillespie, D. S., 1989. Saprolegniasis in a zoo collection of aquatic amphibians. P. 43 in *Proceedings of the 3rd International Colloqium on the Pathology of Reptiles and Amphibians*, Orlando.

Gambier, H., 1924. Sur un Protiste parasite et pathogene des Tritons: *Hepatosphera molgarum*

n. g., n. sp. *Comptes Rendus des Séances de la Societe de Biologie et de des Filiales* **90**: 439–441.

Garcia-Solache, M. A., and Casadevall, A., 2010. Hypothesis: global warming will bring new fungal diseases for mammals. *Mbio* 1:e00061–10.

Garner, T. W. J., Pearman, P. B., Cunningham, A. A. and Fisher, M. C., 2004. Population genetics and disease threats across the entire range of *Rana latastei*. V° Congresso Nazionale della Societas Herpetologica Italica. Calci, Pisa. P. 62.

Garner, T. W. J., Walker, S., Bosch, J., Hyatt, A. D., Cunningham, A. A. and Fisher, M. C., 2005. Chytrid fungus in Europe. *Emerging Infectious Diseases* **11**: 1639–1641.

Garner, T. W. J., Perkins, M. W., Govindarajulu, P., Seglie, D., Walker, S., Cunningham, A. A. and Fisher, M. C., 2006. The emerging amphibian pathogen *Batrachochytrium dendrobatidis* globally infects introduced populations of the North American bullfrog, *Rana catesbeiana*. *Biology Letters* **2**: 455–459.

Garner, T. W. J., Walker, S., Bosch, J., Leech, S., Rowcliffe, J. M., Cunningham, A. A. and Fisher, M. C., 2009. Life history tradeoffs influence mortality associated with the amphibian pathogen *Batrachochytrium dendrobatidis*. *Oikos* **118**: 783–791.

Garner T. W. J., Rowcliffe, J. M. and Fisher, M. C., 2011. Climate, chytridiomycosis or condition: an experimental test of amphibian survival. *Global Change Biology* **17**: 667–675.

Garner T. W. J. Briggs C. J. Bielby J., Fisher, M. C., 2012. Determining when parasites of amphibians are conservation threats to their hosts: methods and perspectives. Pp. 521–538 in *New Directions in Conservation Medicine: Applied Cases of Ecological Health*, ed by A. Aguirre, R. Ostfeld, and P. Daszak. Oxford University Press, Oxford.

Gendrault, J.- L., Steffan, A.- M., Bingen, A. and Kirn, A., 1981. Penetration and uncoating of frog virus 3 in cultured rat Kupffer cells. *Virology* **112**: 375–384.

Gloroiso, J. C., Amborski, R. L., Amborski, G. F. and Culley, D. D., 1974. Microbiological studies on septicaemic bullfrogs (*Rana catesbeiana*). *American Journal of Veterinary Research* **35**: 1241–1245.

Goater, C. P., 1992. Experimental population dynamics of *Rhabdias bufonis* (Nematoda) in toads (*Bufo bufo*): density-dependence in the primary infection. *Parasitology* **104**: 179–187.

Goater, C. P. and Ward, P. I., 1994. Negative effects of *Rhabdias bufonis* (Nematoda) on the growth and survival of toads (*Bufo bufo*). *Oecologia* **89**: 161–165.

Goater, C. P., Semlitsch, R. D. and Bernasconi, M. V., 1993. Effects of body size and parasite infection on the locomotory performance of juvenile toads, *Bufo bufo*. *Oikos* **66**: 129–136.

González-Hernández, M., Denoël, M., Duffus, A. J. L., Garner, T. W. J., Cunningham, A. A. and Acevedo-Whitehouse, K., 2010. Dermocystid infection and associated skin lesions in free-living palmate newts (*Lissotriton helveticus*) from Southern France. *Parasitology International* **59**: 344–350.

Goorha, R., Murti, G., Granoff, A. and Tirey, R. 1978. Macromolecular synthesis in cells infected by frog virus 3: VIII. The nucleus is a site of frog virus 3 DNA and RNA synthesis. *Virology* **84**: 32–50.

Goorha, R. and Granoff, A., 1979. Icosahedral cytoplasmic deoxyriboviruses. Pp. 347–399 in *Comprehensive Virology*, ed by H. Fraenkel-Conrat and R. R. Wagner. Plenum Press, New York.

Granoff, A., 1983. Amphibian herpesviruses. Pp. 367–369 in *The Herpesviruses*, ed by B. Roizmann. Plenum Press, New York.

Gray, M. J., Miller, D. L. and Hoverman, J. T., 2009. Ecology and pathology of amphibian ranaviruses. *Diseases of Aquatic Organisms* **87**: 243–266.

Gray, M. J., Smith, L. M., Miller, D. L. and Bursey, C. R., 2007. Influences of agricultural land use on *Clinostomum attentuatum* metacercariae prevalence in southern Great Plains amphibians, USA *Herpetological Conservation and Biology* **2**: 23–28.

Green, D. E., 2001. Pathology of Amphibia. Pp. 401–485 in *Amphibian Medicine and Captive*

Husbandry, ed by K. M. Wright and B. R. Baker. Krieger Publishing Company, Malabar.

Green, D. E. and Converse, K. A., 2005. Diseases of amphibian eggs and embryos. Pp. 63–71 in *Wildlife Diseases: Landscape Epidemiology, Spatial Distribution and Utilization of Remote Sensing Technology,* ed by S. K. Majumdar, J. E. Huffman, F. J. Brenner and A. I. Panah. The Pennsylvania Academy of Science, Pittsburgh.

Green, D. E. and Harshbarger, J. C., 2001. Spontaneous neoplasia in Amphibia. Pp. 335–400 in *Amphibian Medicine and Captive Husbandry,* ed by K. M. Wright and B. R. Baker. Krieger Publishing Company, Malabar.

Green, D. E. and Sherman, C. K., 2001. Diagnostic histological findings in Yosemite toads (*Bufo canorus*) from a die-off in the 1970's. *Journal of Herpetology* **353**: 92–103.

Green, D. E., Converse, K. A. and Schrader, K. A., 2002. Epizootiology of sixty-four amphibian morbidity and mortality events in the USA, 1996–2001. *Annals of the New York Academy of Science* **969**: 323–339.

Greer, A. L., Briggs, C. J. and Collins, J. P., 2008. Testing a key assumption of host-pathogen theory: density and disease transmission. *Oikos* **117**: 1667–1673.

Guyénot, E. and Naville, A., 1922. Un nouveau protiste du genre *Dermocystidium* parasite de la Grenouille *Dermocystidium ranae* nov. spec. *Revue Suisse de Zoologie* **29**: 133–145.

Hardham, A. R., 2005. *Phytophthora cinnamomi. Molecular Plant Pathology* **6**: 589–604.

Harris, R. N., James, T. Y., Lauer, A., Simon, M. A. and Patel, A., 2006. Amphibian pathogen *Batrachochytrium dendrobatidis* is inhibited by the cutaneous bacteria of amphibian species. *Ecohealth* **3**: 53–56.

Heatwole, H. 2013. Worldwide decline and extinction of amphibians. Chapter 18 (Pp. 259–278 in *The Balance of Nature and Human Impact,* ed by Klaus Rohde. Cambridge University Press. Cambridge. [in press]

Hemingway, V., Brunner, J., Speare, R., Berger, L., 2009. Viral and bacterial diseases of amphibians. Pp. 2963–2985 in *Amphibian Biology, Volume 8 - Amphibian Decline: Diseases, Parasites, Maladies and Pollution,* ed by H. Heatwole and J. W. Wilkinson. Surrey Beatty and Sons, Baulkham Hills.

Herman, R. L., 1984. *Ichthyophonis*-like infection in newts (*Notophthalmus viridescens* Rafinesque). *Journal of Wildlife Diseases* **20**: 55–56.

Hird, D. W., Diesch, S. L., McKinnel, R. G., Gorham, E., Martin, F. B., Kurtz, S. W. and Dubovolny, C., 1981. *Aeromonas hydrophila* in wild-caught frogs and tadpoles (*Rana pipiens*) in Minnesota. *Laboratory Animal Science* **31**: 166–169.

Hoeck, P. and Garner, T. W. J., 2007. Female alpine newts (*Triturus alpestris*) mate first with males signalling fertility benefits. *Biological Journal of the Linnean Society* **91**: 483–491.

Holland, M. P., Skelly, D. K., Kashgarian, M., Bolden, S. R., Harrison, L. M. and Cappello, M., 2007. Echinostome infection in green frogs (*Rana clamitans*) is stage and age dependent. *Journal of Zoology* **271**: 455–462.

Holopainen, R., Ohlemeyer, S., Schütze, H., Bergmann, S. M. and Tapiovaara, H., 2009. Ranavirus phylogeny and differentiation based on major capsid protein, DNA polymerase and neurofilament triplet H1-like protein genes. *Diseases of Aquatic Organisms* **85**: 81–91.

Houlahan, J. E., Findlay, C. S., Schmidt, B. R., Meyer, A. H. and Kuzmin, S. L., 2000. Quantitative evidence for global amphibian population declines. *Nature* **404**: 752–755.

Hoverman, J. T., Gray, M. J. and Miller, D. L., 2010. Anuran susceptibilities to ranaviruses: role of species identity, exposure route, and a novel virus isolate. *Diseases of Aquatic Organisms* **89**: 97–107.

Hulvey, J. P., Padgett, D. E. and Bailey, J. C., 2007. Species boundaries within Saprolegnia (Saprolegniales, Oomycota) based on morphological and DNA sequence data. *Mycologia* **99**: 421–429.

Hunter, B. R., Carlson, D. L., Seppanen, E. D., Killian, P. S., McKinnell, B. K. and McKinnell,

R. G., 1989. Are renal carcinomas increasing in *Rana pipiens* after a decade of reduced prevalence? *American Midland Naturalist* **122**: 307–312.

Hyatt, A. D., Gould, A. R., Zupanovic, Z., Cunningham, A. A., Hengstberger, S., Whittington, R. J., Kattenbelt, J. and Coupar, B. E. H., 2000. Comparative studies of piscine and amphibian iridoviruses. *Archives of Virology* **145**: 301–331.

Hyatt, A. D., Boyle, D. G., Olsen, V., Boyle, D. B., Berger, L., Obendorf, D., Dalton, A., Kriger, K., Hero, J.-M., Hines, H., Phillott, R., Campbell, R., Marantelli, G., Gleason, F. and Colling, A., 2007. Diagnostic assays and sampling protocols for the detection of *Batrachochytrium dendrobatidis*. *Diseases of Aquatic Organisms* **73**: 175–192.

Isaac, N. J., and Cowlishaw, G., 2004. How species respond to multiple extinction threats. *Proceedings of the Royal Society of London, Series B* **271**: 1135–1141.

Isaac, N. J. B., Turvey, S. T., Collen, B., Waterman, C. and Baillie, J. E. M., 2007. Mammals on the EDGE: conservation priorities based on threat and phylogeny. *PLoS ONE* **3**: e296.

James, T. Y., Kauff, F., Schoch, C. L., Matheny, P. B., Hofstetter, V., Cox, C. J., Celio, G., Gueidan, C., Fraker, E., Miadlikowska, J., Lumbsch, H. T., Rauhut, A., Reeb, V., Arnold, A. E., Amtoft, A., Stajich, J. E., Hosaka, K., Sung, G. H., Johnson, D., O'Rourke, B., Crockett, M., Binder, M., Curtis, J. M., Slot, J. C., Wang, Z., Wilson, A. W., Schussler, A., Longcore, J. E., O'Donnell, K., Mozley-Standridge, S., Porter, D., Letcher, P. M., Powell, M. J., Taylor, J. W., White, M. M., Griffith, G. W., Davies, D. R., Humber, R. A., Morton, J. B., Sugiyama, J., Rossman, A. Y., Rogers, J. D., Pfister, D. H., Hewitt, D., Hansen, K., Hambleton, S., Shoemaker, R. A., Kohlmeyer, J., Volkmann-Kohlmeyer, B., Spotts, R. A., Serdani, M., Crous, P. W., Hughes, K. W., Matsuura, K., Langer, E., Langer, G., Untereiner, W. A., Lucking, R., Budel, B., Geiser, D. M., Aptroot, A., Diederich, P., Schmitt, I., Schultz, M., Yahr, R.,

Hibbett, D. S., Lutzoni, F., McLaughlin, D. J., Spatafora, J. W. and Vilgalys, R., 2006. Reconstructing the early evolution of Fungi using a six-gene phylogeny. *Nature* **443**: 818–822.

Jancovich, J. K., Davidson, E. W., Morado, J. F., Jacobs, B. L. and Collins, J. P., 1997. Isolation of a lethal virus from the endangered tiger salamander *Ambystoma tigrinum stebbinsi*. *Diseases of Aquatic Organisms* **31**: 161–167

Jancovich, J. K., Mao, J., Chinchar, V. G., Wyatt, C., Case, S. T., Kumar, S., Valente, G., Subramanian, S., Davidson, E. W., Collins, J. P. and Jacobs, B. L., 2003. Genomic sequence of a ranavirus (family *Iridoviridae*) associated with salamander mortalities in North America. *Virology* **316**: 90–103.

Jancovich, J. K., Davidson, E. W., Parameswaran, N., Mao, J., Chichar, V. G., Collins, J. P., Jacobs, B. L. and Storfer, A., 2005. Evidence for emergence of an amphibian iridoviral disease because of human-enhanced spread. *Molecular Ecology* **14**: 213–224.

Jay, J. M. and Pohley, W. J., 1981. *Dermosporidium penneri* sp. n. from the skin of the American toad, *Bufo americanus* (Amphibia: Bufonidae). *The Journal of Parasitology* **67**: 108–110.

Jirků, M., Fiala, I. and Modrý, D., 2007. Tracing the genus *Sphaerospora*: rediscovery, redescription and phylogeny of the *Sphaerospora ranae* (Morelle, 1929) n. comb. (Myxosporeal Sphaerosporidae), with emendation of the genus *Sphaerospora*. *Parasitology* **134**: 1727–1739.

Jirků, M., Valigurova, A., Koudela, B., Křížek, J., Modrý, D. and Šlapeta, J., 2008. New species of *Cryptosporidium* Tyzzer, 1907 (Apicomplexa) from amphibian host: morphology, biology and phylogeny. *Folia Parasitologica* **55**: 81–94.

Jirků, M., Jirků, M., Oborník, M., Lukeš, J. and Modrý, D., 2009a. *Goussia* Labbé, 1986 (Apicomplexa, Eimeriorina) in Amphibia: diversity, biology, molecular phylogeny and comments on the status of the genus. *Protist* **160**: 123–136.

Jirků, M., Jirků, M., Oborník, M., Lukeš, J. and Modrý, D., 2009b. A model for taxonomic work on homoxenous Coccidia: resdescription, host specificity, and molecular phylogeny of *Eimeria ranae* Dobell, 1909, with a review of anuran-host *Eimeria* (Apicomplexa: Eimeriorina). *Journal of Eukaryotic Microbiology* **56**: 39–51.

Johnson, A. J. and Wellehan, J. F. X., 2005. Amphibian virology. *Veterinary Clinics Exotic Animal Practice* **8**: 53–65.

Johnson, M. L. and Speare, R., 2003. Survival of *Batrachochytrium dendrobatidis* in water: quarantine and disease control implications. *Emerging Infectious Diseases* **9**: 922–925.

Johnson, P. T. J. and Bowerman, J., 2010. Do predators cause frog deformities? The need for an eco-epidemiological approach. *Journal of Experimental Zoology* **314B**: 515–518.

Johnson, P. T. J. and Chase, J. M., 2004. Parasites in the food web: linking amphibian malformations and aquatic eutrophication. *Ecology Letters* **7**: 521–526.

Johnson, P. T. J., Lunde, K. B., Ritchie, E. G. and Launer, A. E., 1999. The effect of trematode infection on amphibian limb development and survivorship. *Science* **284**: 802–804.

Johnson, P. T. J., Lunde, K. B., Zelmer, D. A. and Werner, J. K., 2003. Limb deformities as an emerging parasitic disease in amphibians: evidence from museum specimens and resurvey data. *Conservation Biology* **17**: 1724–1737.

Johnson, P. T. J., Preu, E. R., Sutherland, D. R., Romansic, J. M., Han, B. and Blaustein, A. R., 2006. Adding infection to injury: synergistic effects of predation and parasitism on amphibian malformations. *Ecology* **87**: 2227–2235.

Johnson, P. T. J., Chase, J. M., Dosch, K. L., Hartson, R. B., Gross, J. A., Larson, D. J., Sutherland, D. R. and Carpenter, S. R., 2007. Aquatic eutrophication promotes pathogenic infection in amphibians. *Proceedings of the National Academy of Sciences of the USA* **104**: 15781–15786.

Johnson, P. T. J., Hartson, R. B., Larson, D. J. and Sutherland, D. R., 2008. Diversity and disease: community structure drives parasite transmission and host fitness. *Ecology Letters* **11**: 1017–1026.

Jones, K. E., Patel, N. G., Levy, M. A., Storeygard, A., Balk, D., Gittleman, J. L. and Daszak, P., 2008. Global trends in emerging infectious diseases. *Nature* **451**: 990–994.

Jones, M. D. M., Forn, I., Gadelha, C., Egan, M. J., Bass, D., Massana, R. and Richards, T. A., 2011. Discovery of novel intermediate forms redefines the fungal tree of life. *Nature* **474**: 200–203.

Kamoun, S. 2003. Molecular genetics of pathogenic Oomycetes. *Eukaryotic Cell* **2**: 191–199.

Kang, K. I., Torres-Velez, F. J., Zhang, J., Moore, P. A., Moore, D. P., Rivera, S. and Brown, C. C., 2008. Localization of fibropapilloma-associated turtle herpesvirus in green turtles (*Chelonia mydas*) by *in-situ* hybridization. *Journal of Comparative Pathology* **139**: 218–225.

Kerby, J. and Storfer, A., 2009. Combined effects of atrazine and chlorpyrifos on susceptibility of the Tiger Salamander to *Ambystoma tigrinum* virus. *Ecohealth* **6**: 91–98.

Kerby, J., Hart, A. and Storfer, A., 2011. Combined effects of virus, pesticide, and predator cue on the larval Tiger Salamander (*Ambystoma tigrinum*). *Ecohealth* doi: 10.1007/s10393–011–0682–1.

Kiesecker, J. M., and Blaustein, A. R., 1995. Synergism between UV-B radiation and a pathogen magnifies amphibian embryo mortality in nature. *Proceedings of the National Academy of Sciences of the USA* **92**: 11049–11052.

Kiesecker, J. M., and Blaustein, A. R., 1997. Influences of egg laying behavior on pathogenic infection of amphibian eggs. *Conservation Biology* **11**: 214–220.

Kik, M., Martel, A., Spitzen-van der Sluijs, A., Pasmans, F., Wohlsein, P., Gröne, A. and Rijks, J. M., 2011. Ranavirus-associated mass mortality in wild amphibians, the Netherlands, 2010. A first report. *The Veterinary Journal*, in press.

Kostiukov, M. A., Alekseev, A. N., Bulychev, V. P. and Gordeeva, Z. E., 1986. Experimental evidence for infection of *Culex pipiens* L. mosquitoes by West Nioe fever virus from *Rana ridibunda* Pallas and its transmission by bites. *Meditsinskaia Parazitologiia i Parazitarnye Bolezni (Moskva)* **6**: 76–78.

Kozuch, O., Labuda, M. and Nosek, J., 1978. Isolation of sindbis virus from the frog *Rana ridibunda*. *Acta Virologica* **22**: 78.

Kunst, L. and Valpotic, I., 1968. Nova zarazna bolest zaba uzrokovana virusom. *Veterinarski arhiv* **38**: 108–113.

Kuzmin, Y., Tkach, V. V. and Snyder, S. D., 2003. The nematode genus *Rhabdias* (Nematoda: Rhabdiasidae) from amphibians and reptiles of the Nearctic. *Comparative Parasitology* **70**: 101–114.

Lam, B. A., Walke, J. B., Vredenburg, V. T. and Harris, R. N., 2010. Proportion of individuals with anti-*Batrachochytrium dendrobatidis* skin bacteria is associated with population persistence in the frog *Rana muscosa*. *Biological Conservation* **143**: 529–531.

Lannoo, M., 2008. *Malformed Frogs*. University of California Press, Berkeley.

Leeuwenhoek, A., 1706. Letter concerning the worms in sheep's livers, gnats and animalcula in the excrement of frogs. *Philosophical Transactions of the Royal Society* **22**: 509.

Lefcort, H., Hankcock, K. A., Maur, K. M. and Rostal, D. C., 1997. The effects of used motor oil, silt, and the water mold *Saprolegnia parasitica* on the growth and survival of mole salamanders (Genus *Ambystoma*). *Archives of Environmental Contamination and Toxicology* **32**: 383–388.

Lips, K. R., 1999. Mass mortality and population declines of anurans at an upland site in western Panama. *Conservation Biology* **13**: 117–125.

Lips, K. R., Brem, F., Brenes, R., Reeve, J. D., Alford, R. A., Voyles, J., Carey, C., Livo, L., Pessier, A. P., and Collins, J. P., 2006. Emerging infectious disease and the loss of biodiversity in a Neotropical amphibian community. *Proceedings of the National Academy of Sciences of the USA* **102**: 3165–3170.

Lips, K. R., Diffendorfer, J., Mendelson, J. R. and Sears, M. W., 2008. Riding the wave: reconcil-

ing the roles of disease and climate change in amphibian declines. *PLoS Biology* **6**: e72.

Longcore, J. E., Pessier, A. P. and Nichols, D. K., 1999. *Batrachochytrium dendrobatidis* gen et sp nov, a chytrid pathogenic to amphibians. *Mycologia* **91**: 219–227.

Lötters, S., Kielgast, J., Bielby, J., Schmidtlein, S., Bosch, J., Veith, M., Walker, S. F., Fisher, M. C. and Rödder, D., 2009. The link between rapid enigmatic amphibian decline and the globally emerging chytrid fungus. *Ecohealth* **6**: 358–372.

Lucké, B., 1934. A neoplastic disease of the kidney of the frog, *Rana pipiens*. *American Journal of Cancer* **20**: 352–379.

Lucké, B., 1938. Carcinoma in the leopard frog: its possible causation by a virus. *The Journal of Experimental Medicine* **68**: 457–468.

Lunger, P. D., Darlington, R. W. and Granoff, A., 1965. Cell-virus relationships in the Lucké renal adenocarcinoma: an ultrastructure study. *Annals of the New York Academy of Sciences* **186**: 289.

Luquet, E., Garner, T. W. J., Léna, J.-P., Bruel, C., Joly, P., Lengagne, T., Grolet, O., Plénet, S., 2012. Genetic erosion in wild populations makes resistance to a pathogen more costly. *Evolution* online doi:10.1111/j.1558-5646.2011.01570.x

Majji, S., LaPatra, S., Long S. M., Bryan, L., Sample, R., Sinning, A. and Chinchar, V. G., 2006. *Rana catesbeiana* virus Z (RCV-Z): a novel pathogenic ranavirus. *Diseases of Aquatic Organisms* **73**: 1–11.

Maniero, G. D., Morales, H., Gantress, J. and Robert, J., 2006. Generation of a long-lasting, protective, and neutralizing antibody response to the ranavirus FV3 by the frog *Xenopus*. *Developmental and Comparative Immunology* **30**: 649–657.

Mao, J., Green, D. E., Fellers, G. and Chinchar, V. G., 1999. Molecular characterization of iridoviruses isolated from sympatric amphibians and fish. *Virus Research* **63**: 45–52.

Marcum, R. D., St-Hilaire, S., Murphy, P. J. and Rodnick, K. J., 2010. Effects of *Batrachochytrium dendrobatidis* infection on ion concentrations in the boreal toad *Anaxyrus (Bufo) boreas boreas*. *Diseases of Aquatic Organisms* **91**: 17–21.

Marlow, P. B. and Mizell, S., 1972. Incidence of Lucké renal adenocarcinoma in *Rana pipiens* as determined by histological examination. *Journal of the National Cancer Institute* **48**: 823–829.

Márquez, R., Olmo, J. L. and Bosch, J., 1995. Recurrent mass mortality of larval midwife toads *Alytes obstetricans* in a lake in the Pyrenean Mountains. *Herpetological Journal* **5**: 287–289.

Martel, A., Adriaensen, C., Bogaerts, S., Ducatelle, R., Favoreel, H., Crameri, S., Hyatt, A. D., Haesebrouck, F. and Pasmans, F., 2012. Novel *Chlamydiaceae* disease in captive salamanders. *Emerging Infectious Diseases* **18**: 1020–1022.

Martínez-Solano, I., Bosch, J. and García-París, M., 2003. Demographic trends and community stability in a montane amphibian assemblage. *Conservation Biology* **17**: 238–244.

McAllister C. T., Bursey, C. R., Crawford, J. A., Kuhns, A. R., Shaffer, C. and Trauth, S. E., 2010. Metacercariae of *Clinostomum* (Trematoda: Digenea) from three species of *Ambystoma* (Caudata: Ambystomatidae) from Arkansas and Illinois, USA *Comparative Parasitology* **77**: 25–30.

McCallum, M. L., 2007. Amphibian decline or extinction? Current declines dwarf background extinction rate. *Journal of Herpetology* **41**: 483–491.

McKenzie, V. J., 2007. Human land use and patterns of parasitism in tropical amphibian hosts. *Biological Conservation* **137**: 102–116.

McKinnell, R. G., 1973. The Lucké frog kidney tumor and its herpesvirus. *American Zoologist* **13**: 97–114.

McKinnell, R. G. and DuPlantier, D. P., 1970. Are there renal adenocarcinoma-free populations of leopard frogs? *Cancer Res.* **30**: 2730–2735.

McKinnell, R. G. and John, J. C., 1995. An unexpectedly high prevalence of spontaneous renal carcinoma found in *Rana pipiens* obtained from northern Vermont, USA *Proceedings of the 5th International Colloquim on the Pathology of Reptiles and Amphibians*. Pp. 279–280.

McKinnell, R. G. and Tarin, D., 1984. Temperature-dependent metastasis of the Lucké renal carcinoma and its significance for studies on mechanisms of metastasis. *Cancer Metastasis Reviews* **3**: 373–386.

McKinnell, R. G., Seppanen, E. D., Lust, J. M., Carlson, D. K. and Hunter, B. R., 1989. Lucké renal adenocarcinoma: cell of origin, characterization of malignancy, and genomic potential. *Proceedings of the 3rd International Colloquium on the Pathology of Reptiles and Amphibians*. Pp. 72–73.

Mendoza, L., Taylor, J.W. and Ajello, L., 2002. The class *Mezomycetozoa*: a heterogeneous group of microorganisms at the animal-fungal boundary. *Annual Review of Microbiology* **56**: 314–344.

Mettenleiter, T. C., Keil, G. M. and Fuchs, W., 2008. Molecular biology of animal herpesviruses. Pp. 375–455 in *Animal Viruses: Molecular Biology*, ed by T. C. Mettenleiter and F. Sobrino. Caister Academic Press, Norfolk.

Mikaelian, I., Ouellet, M., Pauli, B., Rodrigue, J., Harshbarger, J. C. and Green, D. M., 2000. *Ichthyophonus*-like infection in wild amphibians from Québec, Canada. *Diseases of Aquatic Organisms* **40**: 195–201.

Miller, D. L., Bursey, C. R., Gray, M. J. and Smith, L. M., 2004. Metacercariae of *Clinostomum attenuatum* in *Ambystoma tigrinum mavortium*, *Bufo cognatus* and *Spea multiplicata* from west Texas. *Journal of Helminthology* **78**: 373–376.

Min, M. S., Yang, S. Y., Bonett, R. M., Vieites, D. R., Brandon, R. A. and Wake, D. B., 2005. Discovery of the first Asian plethodontid salamander. *Nature* **435**: 87–90.

Mişcalencu, D., Alfy, M. E., Mailat, F. and Mihanscu, G. R., 1981. Viral particles in the hepatocytes of *Rana esculenta* (L.). *Revue Roumaine de Médecine-Virologie* **32**: 123–125.

Mitchell, C. M., Hutton, S., Myers, G. S. A., Brunham, R. and Timms, P., 2010. *Chlamydia pneumoniae* is genetically diverse in animals and appears to have crossed the host barrier to humans on (at least) two occasions. *Plos Pathogens* **6**: e1000903.

Mitchell, K. M., Churcher, T. S., Garner, T. W. J. and Fisher, M. C., 2008. Persistence of the emerging infectious pathogen *Batrachochytrium dendrobatidis* outside the amphibian host greatly increases the probability of host extinction. *Proceedings of the Royal Society of London, Series B* **275**: 329–334.

Moral, H., 1913. Über das auftreten von *Dermocystidium pusula* (Pérez), einem einzelligen parasiten der haut des molches bei *Triton cristatus*. *Archiv für Mikroskopische Anatomie* **81**: 381–393.

Morales, H. D. and Robert, J., 2007. Characterization of primary and memory CD8 cell responses against ranavirus (FV3) in *Xenopus laevis*. *Journal of Virology* **81**: 2240–2248.

Morales, H. D., Abramovitz, L., Gertz, J., Sowa, J., Vogel, A. and Robert, J., 2010. Innate immune responses and permissiveness to ranavirus infection of peritoneal leukocytes in the frog *Xenopus laevis*. *Journal of Virology* **84**: 4912–4922.

Moravec, F. and Škoríková, B., 1998. Amphibians and larvae of aquatic insects as new paratenic hosts of *Anguillicola crassus* (Nematoda: Dracunculoidea), a swimbladder parasite of eels. *Diseases of Aquatic Organisms* **34**: 217–222.

Moss, A. S., Carty, N. and Francisco, M. J. S., 2010. Identification and partial characterization of an elastolytic protease in the amphibian pathogen *Batrachochytrium dendrobatidis*. *Diseases of Aquatic Organisms* **92**: 149–158.

Moss, A. S., Reddy, N. S., Dorta, J. I. M. and Francisco, M. J. S., 2008. Chemotaxis of the amphibian pathogen *Batrachochytrium dendrobatidis* and its response to a variety of attractants. *Mycologia* **100**: 1–5.

Mulcare, D. J., 1969. Non-specific transmission of Lucké tumor. Pp. 240–253 in *Biology of Amphibian Tumors*, ed by M. Mizell. Springer-Verlag, New York.

Murphy, P. J., St.-Hilaire, S. and Corn, P. S., 2011. Temperature, hydric environment, and prior pathogen exposure alter the experimental severity of chytridiomycosis in boreal toads. *Diseases of Aquatic Organisms* **95**: 31–42.

Muths, E., Corn, P. S., Pessier, A. P. and Green, D. E., 2003. Evidence for disease-related amphibian decline in Colorado. *Biological Conservation* **110**: 357–365.

Mutschmann, F., 1998. Detection of *Chlamydia psittaci* in amphibians using an immunofluorescence test (IFT). *Berliner und Münchener Tierärztliche Wochenschrift* **111**: 187–189.

Mutschmann, F., 2007. Chytridiomycosis in Germany - an overview. *Proceedings of the 7th International Symposium on Pathology and Medicine in Reptiles and Amphibians (Berlin 2004)*.

Mutschmann, F. and Schneeweiss, D., 2008. Herpes Virus Infektionen bei *Pelobates fuscus* und anderen Anuren im Berlin-Brandenburger Raum. *Rana. Sonderheft* **5**: 113–118.

Nieto, N. C., Camann, M. A., Foley, J. E. and Reiss, J. O., 2007. Disease associated with integumentary and cloacal parasites in tadpoles of northern red-legged frog *Rana aurora aurora*. *Diseases of Aquatic Organisms* **78**: 61–71.

Oevermann, A., Schmidt-Posthaus, H., Hoeck, P. and Nadia, R., 2005. *Amphibiocystidium*-like protozoans causing generalized infection in a laboratory colony of alpine newts. *Proceedings of the 23rd Meeting of the European Society of Veterinary Pathology*. P. 152.

Ohst, T., Gräser, Y., Mutschmann, F. and Plötner, J., 2011. Neue Erkentnisse zur Gefährdung europäischer Amphibien durch den Hautpilz *Batrachochytrium dendrobatidis*. *Zeitschrift für Feldherpetologie* **18**: 1–17.

Olgun, K., Arntzen, J. W., Kuzmin, S., Papenfuss, T., Ugurtas, I., Tarkhnishvili, D., Sparreboom, M., Anderson, S., Turiyev, B., Ananjeva, N., Kaska, Y., Kumlutaş, Y., Avci, A., Üzüm, N. and Kaya, U., 2010. *Ommatotriton ophryticus*. *In*: IUCN 2010. IUCN Red List of Threatened Species. Version 2010.4.

Orizaola, G. and Braña, F., 2006. Effect of salmonid introduction and other environmental characteristics on amphibian distribution and abundance in mountain lakes of northern Spain. *Animal Conservation* **9**: 171–178.

Orton, F. and Routledge, E., 2011. Agricultural intensity *in ovo* affects growth, metamorphic development and sexual differentiation in the common toad (*Bufo bufo*). *Ecotoxicology* **20**: 901–911.

Parker, B. J., Barribeau, S. M., Laughton, A. M., de Roode, J. C. and Gerardo, N. M., 2011. Non-immunological defense in an evolutionary framework. *Trends in Ecology and Evolution* **26**: 242–248.

Pascolini, R., Daszak, P., Cunningham, A.A., Tei, S., Vagnetti, D., Bucci, S., Fagotti, A. and Di Rosa, I., 2003. Parasitism by *Dermocystidium ranae* in a population of *Rana esculenta* complex in Central Italy and description of *Amphibiocystidium* n. gen. *Diseases of Aquatic Organisms* **56**: 65–74.

Pasmans, F., Muijsers, M., Maes, S., Van Rooij, P., Brutyn, M., Ducatelle, R., Haesebrouck, F. and Martel, A., 2010. Chytridiomycosis related mortality in a midwife toad (*Alytes obstetricans*) in Belgium. *Flemish Veterinary Journal* **79**: 460–462.

Pearman, P. B. and Garner, T. W. J., 2005. Susceptibility of Italian agile frog populations to an emerging strain of Ranavirus parallels population genetic diversity. *Ecology Letters* **8**: 401–408.

Pearman, P. B., Garner, T. W. J., Straub, M. and Greber, U. F., 2004. Response of *Rana latastei* to the Ranavirus FV3: a model for viral emergence in a naïve population. *Journal of Wildlife Disease* **40**: 600–609.

Pérez, C., 1907. *Dermocystis pusula* organisme nouveau parasite de la peau des tritons. *Comtes Rendus des Séances de la Société de Biologie et de ses Filiales* **63**: 445–447.

Pérez, C., 1913. *Dermocystidium pusula*: Parasite de la peau des Tritons. *Archives de Zoologie Expérimentale et Générale* **52**: 343–357.

Pessier, A. P., Nichols, D. K., Longcore, J. E. and Fuller, M. S., 1999. Cutaneous chytridiomycosis in poison dart frogs (*Dendrobates* sp.) and White's tree frogs (*Litoria caerulea*).

Journal of Veterinary Diagnostic Investigation **11**: 194–199.

Pedersen, A. B., Altizer, S., Poss, M., Cunningham, A. A. and Nunn, C. L., 2005. Patterns of host specificity and transmission among parasites of wild primates. *International Journal of Parasitology* **35**: 647–657.

Petrisko, J. E., Pearl, C. A., Pilliod, D. S., Sheridan, P. P., Williams, C. F., Peterson, C. R. and Bury, R. B., 2008. Saprolegniaceae identified on amphibian eggs throughout the Pacific Northwest, USA, by internal transcribed spacer sequences and phylogenetic analysis. *Mycologia* **100**: 171–180.

Picco, A. M., and Collins, J. P., 2008. Amphibian commerce as a likely source of pathogen pollution. *Conservation Biology* **22**: 1582–1589.

Piha, H., Pekkonen, M. and Merilä, J., 2006. Morphological abnormalities in amphibians in agricultural habitats: a case study of the common frog *Rana temporaria*. *Copeia* **2006**: 810–817.

Piotrowski, J. S., Annis, S. L. and Longcore, J. E., 2004. Physiology of *Batrachochytrium dendrobatidis*, a chytrid pathogen of amphibians. *Mycologia* **96**: 9–15.

Poisson, C., 1937. Sur une nouvelle espèce du genre Dermomycoides Granata 1919: *Dermomycoides armoriacus* Poisson 1936 parasite cutané de *Triturus palmatus* (Schneider). Genèse et structure de la zoospore. *Bulletin Biologique de la France et de la Belgique* **71**: 91–116.

Prudhoe, S. and Bray, R. A., 1982. *Platyhelminth Parasites of the Amphibia*. British Museum (Natural History), London.

Puky, M., 2006. Amphibian deformity frequency and monitoring methodology in Hungary. *Froglog* **74**: 3–4.

Rachowicz, L. J. and Briggs, C. J., 2007. Quantifying the disease transmission functions: effects of density on *Batrachochytrium dendrobatidis* transmission in the mountain yellow-legged frog *Rana muscosa*. *Journal of Animal Ecology* **76**: 711–721.

Raffel, T. R., LeGros, R. P., Love, B. C., Rohr, J. R. and Hudson P. J., 2009. Parasite age-intensity relationships in red-spotted newts: does immune memory influence salamander disease dynamics? *International Journal for Parasitology* **39**: 231–241.

Raffel, T. R., Bommarito, T., Barry, D. S., Witiak, S. M. and Shackleton, L. A., 2008. Widespread infection of the Eastern red-spotted newt (*Notophthalmus viridescens*) by a new species of Amphibiocystidium, a genus of fungus-like mesomycetozoan parasites not previously reported in North America. Parasitology **135**: 203–215.

Raffel, T. R., Dillard, J. R. and Hudson P. J., 2006. Field evidence for leech-borne transmission of amphibian *Ichthyophonus* sp. *Journal of Parasitology* **92**: 1256–1264.

Ragan, M. A., Goggin, C. L., Cawthorn, R. J., Cerenius, L., Jamieson, A. V. C., Plourde, S. M., Rand, T. G., Söderhäll, K. and Gutell. R. R., 1996. A novel clade of protistan parasites near the animal–fungal divergence. Proceedings of the National Academy of Science U.S.A. **93**: 11907–11912.

Ramsey, J. P., Reinert, L. K., Harper, L. K., Woodhams, D. C. and Rollins-Smith, L. A., 2010. Immune defenses against *Batrachochytrium dendrobatidis*, a fungus linked to global amphibian declines, in the South African clawed frog, *Xenopus laevis*. *Infection and Immunity* **78**: 3981–3992.

Reading, C. J., 2007. Linking global warming to amphibian declines through its effects on female body condition and survivorship. *Oecologia* **151**: 125–131.

Reed, K. D., Ruth, G. R., Meyer, J. A. and Shukla, S. K., 2000. *Chlamydia pneumonia* infection in a breeding colony of African clawed frogs (*Xenopus tropicalis*). *Emerging Infectious Diseases* **6**: 196–199.

Remy, P., 1931. Présence de *Dermocystidium ranae* (Guyénot et Naville) chez une *Rana esculenta* L. de Lorraine. *Annales de Parasitologie* **9**: 1–3.

Retallick, R. W. R. and Miera, V., 2007. Strain differences in the amphibian chytrid *Batrachochytrium dendrobatidis* and non-permanent, sub-lethal effects of infection. *Diseases of Aquatic Organisms* **75**: 201–207.

Ribas, L., Li, M.-S., Doddington, B., Robert, J., Seidel, J. A., Kroll, J. S., Zimmerman, L., Grassly, N. C., Garner, T. W. J. and Fisher, M. C., 2009. Expression profiling the temperature-dependent amphibian response to infection by *Batrachochytrium dendrobatidis*. *PLoS ONE* **4**: e8408.

Ridenhour, B. J. and Storfer, A. T., 2008. Geographically variable selection in *Ambystoma tigrinum* virus (Iridiviridae) throughout the western USA. *Journal of Evolutionary Biology* **21**: 1151–1159.

Robert, J., 2010. Emerging ranaviral infectious diseases and amphibian decline. *Diversity* **2**: 314–330.

Robert, J., Abramovitz, L., Gantress, J. and Morales, H. D., 2007. *Xenopus laevis*: a possible vector of Ranavirus infection? *Journal of Wildlife Diseases* **43**: 645–652

Robert, V. A., and Casadevall, A., 2009. Vertebrate endothermy restricts most fungi as potential pathogens. *Journal of Infectious Diseases* **200**: 1623–1626.

Robinson, J., Griffiths, R.A. and Jeffries, P., 2003. Susceptibility of frog (*Rana temporaria*) and toad (*Bufo bufo*) eggs to invasion by *Saprolegnia*. *Amphibia-Reptilia* **24**: 261–268.

Robinson, R. A., Lawson, B., Toms, M. P., Peck, K. M., Kirkwood, J. K., Chantrey, J., Clatworthy, I. R., Evans, A. D., Hughes, L. A., Hutchinson, O. C., John, S. K., Pennycott, T. W., Perkins, M. W., Rowley, P. S., Simpson, V. R., Tyler, K. M., Cunningham, A. A., 2010. Emerging infectious disease leads to rapid population declines of common British birds. *PLoS ONE* **5**: e12215.

Rohr, J. R., Raffel, T. R., Sessions, S. K. and Hudson, P. J., 2008a. Understanding the net effects of pesticides on amphibian trematode infections. *Ecological Applications* **18**: 1743–1753.

Rohr, J. R., Schotthoefer, A. M., Raffel, T. R., Carrick, H. J., Halstead, N., Hoverman, J. T., Johnson, C. M., Johnson, L. B., Lieske, C., Piwoni, M. D., Schoff, P. K. and Beasley, V.R., 2008b. Agrochemicals increase trematode infections in a declining amphibian species. *Nature* **455**: 1235–1240.

Rojas, S., Richards, K., Jancovich, J. K. and Davidson, E. W., 2005. Influence of temperature on Ranavirus infection in larval salamanders, *Ambystoma tigrinum*. *Diseases of Aquatic Organisms* **63**: 95–100.

Romansic, J. M., Diez, K. A., Higashi, E. M. and Blaustein, A. R., 2006. Effects of nitrate and the pathogenic water mold *Saprolegnia* on survival of amphibian larvae. *Diseases of Aquatic Organisms* **68**: 235–243.

Romansic, J. M., Higashi, E. M., Diez, K. A. and Bluastein, A. R., 2007. Susceptibility of newly-metamorphosed frogs to a pathogenic water mould (*Saprolegnia* sp.). *Herpetological Journal* **17**: 161–166.

Rothenberg, S., Chinchar, V. G. and Dever, T. E., 2011. Characterization of a ranavirus inhibitor of the antiviral protein kinase PKR. *BMC Microbiology* **11**: 56.

Rowley, J. J. L., Skerratt, L. F., Alford, R. A. and Campbell, R., 2007. Retreat sites of rain forest stream frogs are not a reservoir for *Batrachochytrium dendrobatidis* in northern Queensland, Australia. *Diseases of Aquatic Organisms* **74**: 7–12.

Ruthig, G. R., 2008. The influence of temperature and spatial distribution on the susceptibility of southern leopard frog eggs to disease. *Oecologia* **156**: 895–903.

Sagvik, J., Uller, T., Stenlund, T. and Olsson, M., 2008a. Intraspecific variation in resistance of frog eggs to fungal infection. *Evolutionary Ecology* **22**: 193–201.

Sagvik, J., Uller, T. and Olsson, M., 2008b. A genetic component of resistance to fungal infection in frog embryos. *Proceedings of the Royal Society Series B* **275**: 1393–1396.

St-Amour, V., Wong, W., Garner, T. W. J. and Lesbarrères, D., 2008. Anthropogenic influence on prevalence of two amphibian pathogens. *Emerging Infectious Diseases* **14** 1175–1176.

Sanchez-Morgado, J. M., Gallagher, A., Johnson, L. K., 2009. *Mycobacterium gordonae* infection in a colony of African clawed frogs (*Xenopus tropicalis*). *Laboratory Animals* **43**: 300–303.

Schloegel, L. M., Picco, A. M., Kilpatrick, A. M., Davies, A. J., Hyatt, A. D. and Daszak, P., 2009. Magnitude of the U.S. trade in amphibians and presence of *Batrachochytrium dendrobatidis* and ranavirus infection in imported North American bullfrogs (*Rana catesbeiana*). *Biological Conservation* **142**: 1420–1426.

Schloegel, L. M., Daszak, P., Cunningham, A. A., Speare, R. and Hill, B., 2010. Two amphibian diseases, chytridiomycosis and ranaviral disease, are now globally notifiable to the World Organization for Animal Health (OIE): an assessment. *Diseases of Aquatic Organisms* **92**: 101–108.

Schock, D. M., Bollinger, T. K., Chinchar, V. G., Jancovich, J. K. and Collins, J. P., 2008. Experimental evidence that amphibian ranaviruses are multi-host pathogens. *Copeia* **2008**: 133–143.

Schotthoefer, A. M., Cole, R. A. and Beasley, V. R., 2003a. Relationship of tadpole stage to location of echinostome cercariae encystment and the consequences for tadpole survival. *Journal of Parasitology* **89**: 475–482.

Schotthoefer, A. M., Koehler, A. V., Meteyer, C. U. and Cole, R. A., 2003b. Influence of *Ribeiroia ondatrae* (Trematoda: Digenea) infection on limb development and survival of northern leopard frogs (*Rana pipiens*): effects of host stage and parasite-exposure levels. *Canadian Journal of Zoology* **81**: 1144–1153.

Segev, O., Mangel, M. and Blaustein, L., 2009. Deleterious effects by mosquitofish (*Gambusia affinis*) on the endangered fire salamander (*Salamandra infraimmaculata*). *Animal Conservation* **12**: 29–37.

Seixas, F., da Luz Martins, M., de Lurdes Pinto, M., Travassos, P. J., Miranda, M. and dos Anjos Pires, M., 2008. A case of pulmonary cryptococcosis in a free-living toad (*Bufo bufo*). *Journal of Wildlife Diseases* **44**: 460–463.

Sessions, S. K. and Ruth, S. B., 1990. Explanation for naturally occurring supernumerary limbs in amphibians. *Journal of Experimental Zoology* **254**: 38–47.

Sharifian-Fard, M., Pasmans, F., Adriaensens, C., Devisscher, S., Adriaens, T., Louette, G. and Martel, A., 2011. Ranavirosis in invasive bullfrogs, Belgium. *Emerging Infectious Diseases* **17**: 2371–2372.

Simoncelli, F., Fagotti, A., Dall'Olio, R., Vagnetti, D., Pascolini, R. and Di Rosa, I., 2005. Evidence of *Batrachochytrium dendrobatidis* infection in water frogs of the *Rana esculenta* complex in central Italy. *Ecohealth* **2**: 307–312.

Skerratt, L. F., Berger, L., Speare, R., Cashins, S., McDonald, K. R., Phillott, A. D., Hines, H. B. and Kenyon, N., 2007. Spread of chytridiomycosis has caused the rapid global decline and extinction of frogs. *Ecohealth* **4**: 125–134.

Skerratt, L. F., Garner, T. W. J. and Hyatt, A. D., 2009. Determining causality and controlling disease is based on collaborative research involving multidisciplinary approaches. *Ecohealth* **6**: 331–334.

Soares, C., Alves de Matos, A. P., Arntzen, J. W. Carretero, M. and Loureiro, A., 2002a. Amphibian mortality in a national park in the north of Portugal. *Froglog* **56**: 1.

Soares, C., Alves de Matos, A. P., Loureiro, A. and Carretero, M., 2002b. Episódios de mortalidade nas populações de anfíbios na Lagoa dos Carris - Parque Nacional da Peneda-Gerês (PNPG), Portugal. *VII Congresso Luso-Espanhol de Herpetologia, Évora, Portugal.*

Stagni, G., Dall'Olio, R., Fusini, U., Mazzotti, S., Scoccianti, C. and Serra, A., 2004. Declining populations of Apennine Yellow Bellied Toad *Bombina pachypus* in northern Apennines (Italy): is *Batrachochytrium dendrobatidis* the main cause? *Italian Journal of Zoology* **71**: 5–13.

Stajich, J. E., Berbee, M. L., Blackwell, M., Hibbett, D. S., James, T. Y., Spatafora, J. W. and Taylor, J. W., 2009. The fungi. *Current Biology* **19**: R840-R845.

Stockwell, M., Clulow, J. and Mahoney, M., 2010. Host species determines whether infection load increases beyond disease-causing thresholds following exposure to the amphibian chytrid fungus. *Animal Conservation* **13**: S62–71.

Storfer, A., Alfaro, M. E., Rindenhour, B. J., Jancovich, J. K., Mech, S. G., Parris, M. J. and Collins, J. P., 2007. Phylogenetic concordance analysis shows an emerging pathogen is novel and endemic. *Ecology Letters* **10**: 1075–1083.

Strijbosch, H., 1979. Habitat selection of amphibians during their aquatic phase. *Oikos* **33**: 363–372.

Stuart, S. N., Chanson, J. S., Cox, N. A., Young, B. E., Rodrigues, A. S. L., Fischman, D. L. and Waller, R. W., 2004. Status and trends of amphibian declines and extinctions worldwide. *Science* **306**: 1783–1786.

Suykerbuyk, P., Vleminckx, K., Pasmans, F., Stragier, P., Ablordey, A., Tran, H.T., Hermans, K., Fleetwood, M., Meyers, W.M. and Portaels, F., 2007. *Mycobacterium liflandii* infection in European colony of *Silurana tropicalis. Emerging Infectious Diseases* **13**: 743–746.

Symonds, E. P., Trott, D. J., Bird, P. S. and Mills, P., 2008. Growth characteristics and enzyme activity in *Batrachochytrium dendrobatidis* isolates. *Mycopathologia* **166**: 143–147.

Sztatecsny, M. and Glasner, F., 2011. From the eastern lowlands to the western mountains: first records of the chytrid fungus *Batrachochytrium dendrobatidis* in wild amphibian populations from Austria. *Herpetological Journal* **21**: 87–90.

Taylor, J. W., Jacobson, D. J. and Fisher, M. C., 1999. The evolution of asexual fungi: Reproduction, speciation and classification. *Annual Review of Phytopathology* **37**: 197–246.

Taylor, S. K., 2001. Mycoses. Pp. 181–191 in *Amphibian Medicine and Captive Husbandry*, ed by K. M. Wright and B. R. Baker. Krieger Publishing Company, Malabar.

Teacher, A. F. G., Garner, T. W. J. and Nichols, R. A., 2009. Evidence for directional selection at a novel Major Histocompatibility Class 1 marker in wild common frogs (Rana temporaria) exposed to a viral pathogen (Ranavirus). PLoS Biology 4, e4616. DOI:10.1377/journal.pone.004616

Teacher, A. G. F., Cunningham, A. A. and Garner, T. W. J., 2010. Assessing the long-term impact of *Ranavirus* infection in wild common frog populations. *Animal Conservation* **13**: 514–522.

Tobler, U. and Schmidt, B. R., 2010. Within- and among-population variation in chytridiomycosis-induced mortality in the toad *Alytes obstetricans*. *PLoS ONE* **5**: e10927.

Trott, K. A., Stacy, B. A., Lifland, B. D., Diggs, H. E., Harland R. M., Khokha, M. K., Grammer T. C. and Parker, J. M., 2004. Characterization of a *Mycobacterium ulcerans*-like infection in a colony of African tropical clawed frogs (*Xenopus tropicalis*). *Comparative Medicine* **54**: 309–317.

Twedell, K.S., 1989. Herpesviruses: interaction with frog renal cells. Pp. 13–29 in *Viruses of Lower Vertebrates*, ed by W. Ahne and E. Kurstak. Springer Verlag, Heidelberg.

van der Sluijs, A., Martel, A., Wombwell, E., Van Rooij, P., Zollinger, R., Woeltjes, T., Rendle, M., Haesebrouck, F., and Pasmans, F., 2011. Clinically healthy amphibians in captive collections and at pet fairs: reservoir of *Batrachochytrium dendrobatidis*. *Amphibia Reptilia* **32**: 419–423.

van West, P., Phillips, A. J., Anderson, V. L., Robertson, E. J. and Secombes, C. J., 2008. New insights into animal pathogenic oomycetes. *Trends in Microbiology* **16**: 13–19.

Voyles, J., Berger, L., Young, S., Speare, R., Webb, R., Warner, J., Rudd, D., Campbell, R. and Skerratt, L. F., 2007. Electrolyte depletion and osmotic imbalance in amphibians with chytridiomycosis. *Diseases of Aquatic Organisms* **77**: 113–118.

Voyles, J., Young, S., Berger, L., Campbell, C., Voyles, W. F., Dinudom, A., Cook, D., Webb, R., Alford, R. A., Skerratt, L. F. and Speare, R., 2009. Pathogenesis of chytridiomycosis, a cause of catastrophic amphibian declines. *Science* **326**: 582–586.

Voyles, J., Rosenblum, E. B. and Berger, L., 2011. Interactions between *Batrachochytrium dendrobatidis* and its amphibian hosts: a review of pathogenesis and immunity. *Microbes and Infection* **13**: 25–32.

Vredenburg, V. T., Knapp, R. A., Tunstall, T. S. and Briggs, C. J., 2010. Dynamics of an emerging disease drive large-scale amphibian population extinctions. *Proceedings of the National Academy of Sciences of the USA* **107**: 9689–9694.

Vulpian, 1859. Note sur un nouveau Distome de la grenouille. *Comptes Rendus des Séances et Mémoires de la Société de Biologie, Paris* Year 1858: 150–152.

Wake, D. B. and Vredenburg, V. T., 2008. Are we in the midst of the sixth mass extinction? A view from the world of amphibians. *Proceedings of the National Academy of Sciences of the USA* **105**: 11466–11473.

Walker, C. A. and van West, P., 2007. Zoospore development in the oomycetes. *Fungal Biology Reviews* **21**: 10–18.

Walker, S. F., Baldi Salas, M., Jenkins, D., Garner, T. W. J., Cunningham, A. A., Hyatt, A. D., Bosch, J. and Fisher, M. C., 2007. Environmental detection of *Batrachochytrium dendrobatidis* in a temperate climate. *Diseases of Aquatic Organisms* **77**: 105–112.

Walker, S. F., Bosch, J., James, T. Y., Litvintseva, A. P., Valls, J. A. O., Piña, S., Garcia, G., Rosa, G. A., Cunningham, A. A., Hole, S., Griffiths, R., and Fisher, M. C., 2008. Invasive pathogens threaten species recovery programs. *Current Biology* **18**: R853–R854.

Walker, S. F., Bosch, J., Gomez, V., Garner, T. W. J., Cunningham, A. A., Schmeller, D. S., Ninyerola, M., Henk, D., Ginestet, C., Christian-Philippe, A. and Fisher, M. C., 2010. Factors driving pathogenicity versus prevalence of the amphibian pathogen *Batrachochytrium dendrobatidis* and chytridiomycosis in Iberia. *Ecology Letters* **13**: 372–382.

Walls, S. C. and Jaeger, R. G., 1987. Aggression and exploitation as mechanisms of competition in larval salamanders. *Canadian Journal of Zoology* **65**: 2938–2944.

Willis D. B. and Granoff, A. 1978. Macromolecular synthesis in cells infected by frog virus 3.

IX. Two temporal classes of early viral RNA. *Virology* **86**: 443–453.

Woodhams, D. C., Ardipradja, K., Alford, R. A., Marantelli, G., Reinert, L. K. and Rollins-Smith, L.A., 2007. Resistance to chytridiomycosis varies among amphibian species and is correlated with skin peptide defenses. *Animal Conservation* **10**: 409–417.

Woodhams, D. C., Alford, R. A., Briggs, C. J., Johnson, M. and Rollins-Smith, L. A., 2008. Life-history trade-offs influence disease in changing climates: strategies of an amphibian pathogen. *Ecology* **89**: 1627–1639.

Woodhams, D. C., Bosch, J., Briggs, C. J., Cashins, S., Davis, L. R., Lauer, A., Muths, E., Puschendorf, R., Schmidt, B. R., Sheafor, B. and Voyles, J., 2011. Mitigating amphibian disease: strategies to maintain wild populations and control chytridiomycosis. *Frontiers in Zoology* **8**: 8.

Yildirimhan, H. S., Bursey, C. R. and Goldberg, S. R., 2005. Helminth parasites of the Caucasian salamander, *Mertensiella caucasica*, from Turkey. *Comparative Parasitology* **72**: 75–87.

Zupanovic, Z., Lopez, G., Hyatt, A., Shiell, B. J. and Robinson, A. J., 1998. An improved enzyme linked immunosorbent assay for detection of anti-ranavirus antibodies in the serum of the giant toad (*Bufo marinus*). *Developmental and Comparative Immunology* **22**: 573–585.

32 Conservation and declines of amphibians in Ireland

Ferdia Marnell

Abbreviations and acronyms used in the text or references:

EU	*European Union*
HMSO	*Her Majesty's Stationery Office*
IUCN	*International Union for the Conservation of Nature*
NI	*Northern Ireland*
NPWS	*National Parks and Wildlife Service*
RoI	*Republic of Ireland*
UK	*United Kingdom*

I. Introduction

This chapter covers the island of Ireland, i.e. both Northern Ireland (NI) and the Republic of Ireland (RoI).

The island of Ireland lies at the western edge of Europe. It is characterized by a central plain surrounded by relatively low-lying hills and by mountains nearer the coast. There are many large rivers, several of which retain significant floodplains and associated wetlands. Ireland has a mild Atlantic climate and, while thick woodlands covered the island until the 17th century, today only 10% of the country is forested. Agriculture dominates the Irish landscape and grassland; pasture and silage accounts for as much as 70% of the land-use. The traditional small-field system persists over much of Ireland and, despite extensive mechanization and localized removal of the boundaries of fields, there is still a significant network of hedgerows and lines of trees across the island.

Being at the edge of Europe also explains the limited fauna found on the island. Natural recolonization by animals following the last glaciation was largely limited to cold-tolerant species although more recent human-mediated introductions have also taken place. Today there are only three species of amphibians in Ireland – the smooth newt (*Lissotriton vulgaris*), the common frog (*Rana temporaria*) and the natterjack toad (*Bufo [Epidalea] calamita*). The first two are widespread throughout the island. The natterjack, however, is limited to a dozen or so populations in the southwestern part, with a further introduced population in a sand-dune system in the southeast.

The first Irish atlas that included amphibians was published in 1974 (Crichton 1974) with a second edition five years later (Ní Lamhna 1979). More recent field surveys for newts (Marnell

1998b; O'Neill *et al.* 2004), frogs (Marnell 1999) and toads (Bécart *et al.* 2007) have provided a more complete understanding of the distribution and status of these species in Ireland.

II. The Irish amphibians

A. *Lissotriton vulgaris*

This species is protected under the RoI's Wildlife Act and the NI Wildlife Order. This protection makes it an offense to deliberately capture or kill the animal without an appropriate licence. The newts' breeding sites are also protected although provision is made for a number of exemptions.

The smooth newt can be quite localized in Ireland and is not well known or often recorded because of its tendency to favour weedy ponds and ditches. However, dedicated surveys have found it to occur right across the island and it is still considered widespread (Marnell 1998b; O'Neill *et al.* 2004). Nonetheless, there are concerns about loss of habitat. O'Neill *et al.* (2004) concluded that "Smooth newts are widespread in Northern Ireland and the incidence of newts in suitable water bodies gives little cause for concern". However, suitable habitat for newts appears to be in steep and ongoing decline.

B. *Rana temporaria*

This species is protected under the RoI's Wildlife Act and the NI Wildlife Order. This protection makes it an offense to deliberately capture or kill the animal without an appropriate licence. The frogs' breeding sites are also protected although provision is made for a number of exemptions.

This species of frog is one of Ireland's most common and familiar vertebrates. Nonetheless it is listed on Annex V of the EU's Habitats Directive and the last conservation assessment raised concerns about habitat loss associated with agricultural intensification and urban and suburban spread (NPWS 2007). A full national survey is underway in the Republic of Ireland at present. As well as examining habitat preferences, this survey will also provide an estimate of population size and update the understanding of the species' full range across the Republic of Ireland.

C. *Bufo [Epidalea] calamita*

This species is protected under the RoI's Wildlife Act. This protection makes it an offense to deliberately capture or kill the animal without an appropriate licence. The natterjack's breeding sites are also strictly protected due to its listing on Annex IV of the EU's Habitats Directive.

Beebee (2002) summarized the historical status of this species in Ireland since it was first discovered in the early 1800s and concluded that, before the 1970s, a substantial contraction in its range seemed to have occurred, in particular around Castlemaine Harbour. He further reported that no natterjacks' breeding sites had been lost in Ireland since a survey by Gresson and O'Dubhda in 1974. It appears that prior to the 1970s natterjacks are likely to have disappeared in Ireland from approximately half of their historical range (Beebee 2008). The historical range (for the period 1805-1971) was estimated to have a total area of 188 km2 (more than twice the area of the current range) (NPWS 2007).

More recent data (May and Beebee 2008) imply that the Irish natterjack populations may never have been continuously distributed around Castlemaine Harbour. The construction of extensive seawalls in the 1930s and 1940s and improved land drainage probably removed extensive areas of toad habitat along the southern and eastern coasts of the harbour. However, May and Beebee (2002) concluded that "there have always been substantial barriers to toad movement around the bay, such as the rivers Caragh, Laune and Maine. Much of the northern coast between Roscullen and Inch is drier and less suitable than the south coast as natterjack habitat, and Inch toads may (as the genetic data suggest) have been isolated from others around the harbour for a long time." This information is leading to a reassessment of the former range of the species. Nonetheless, the range has certainly declined and efforts are now underway to reverse that trend (see below).

III. Conservation measures and monitoring programmes

Early conservation work for the natterjack toad included a reintroduction programme at the sand dunes at Caherdaniel in southern Kerry in the 1990s and a translocation attempt at two sites in the southeastern part of the country. About that time there was significant pressure to develop coastal dune systems in Ireland for tourism. In particular, development of golf courses was exempt from planning permission at that time.

There had been old records of natterjacks at Caherdaniel (Dover 1877; MacDougald 1942) but the species appeared to have been extirpated by the second half of the 20th century. The National Parks and Wildlife Service (NPWS) initiated a reintroduction programme in 1991, transferring spawn from Castlegregory where toads were abundant; this project lasted for a number of successive years. About the same time, toadspawn was translocated from Castlegregory to Ballyteigue and to the Raven dune systems in Wexford, southeastern Ireland. These locations were well outside the known range of the toad but the sites were deemed to contain suitable habitat and had the advantage of being in state ownership and were therefore free from pressures to develop. The introduction to Ballyteigue was not a success and the toads never established there. The Raven has proven more successful, although it was necessary to introduce spawn over a 10-year period before a self-sustaining population became established.

In 2002, a conservation plan for natterjacks was commissioned by the NPWS (Beebee 2002). This plan reviewed the documented records of the toad in Ireland and summarized its current status. It went on to identify actions, on a site-by-site basis, that were necessary to ensure the long-term favourable conservation status of the toad in Ireland. These recommendations continue to form the basis for the current conservation strategy for the natterjack toad in Ireland.

Following the publication of Beebee's (2002) report, an *ad hoc* programme of pond creation was funded by the Department of the Environment, with a small number of ponds dug where amenable private landowners could be found within the natterjack's range. During this time, a few ponds were dug or deepened at Fermoyle and Tullaree on the northern coast of the Dingle peninsula and at Roscullen on Castlemaine Harbour. These efforts produced some limited local success (Bécart *et al*. 2007). However, following the Article 17 report to the European Commission in 2007 (NPWS 2007), wherein the natterjack toad was deemed to be in Bad conservation status, a more ambitious programme of pond creation was launched by NPWS. Under this new programme, farmers within areas inhabited by natterjack toads could sign up for a five-year agreement. In year 1 they would be paid to dig two toad ponds; in years 2–5 they would be paid to maintain the ponds, and the surrounding swards, for toads. This scheme proved popular with farmers and to date 48 have joined and almost 100 new ponds have been dug.

An initial three-year monitoring programme ran from 2004 to 2006 and, using population-modelling based on detailed observations in the field, the first population estimates for the toad in Ireland were produced (Bécart *et al*. 2007). A further two-year monitoring programme was initiated in 2011 to survey all newly created ponds, as well as the old traditional ponds, throughout the 2011 and 2012 breeding seasons. Initial indications are that toads have already colonized about 15 of the new ponds (*ca*. 15%) and it is hoped that further ponds will be colonized in the years to come.

Because the frog is listed on the Habitats Directive, member states are obliged to undertake national monitoring programmes and to report on conservation status. A first-ever national frog survey was initiated in the Republic of Ireland in 2010 and another report is due at the end of 2011. This survey will provide the first estimates of the sizes of frog populations in Ireland and will also examine the habitat adaptability of the species. Early indications are that the species is widespread and abundant.

IV. Status of Irish amphibians on the red data list

The first Red Data Book of Irish vertebrates appeared in 1993 (Whilde 1993). The base of knowledge for most species has improved significantly since then and an updated Red List of amphibians, reptiles, and freshwater fish has just been published (King *et al.* 2011). The 2011 assessments used the most recent categories on the IUCN Red Data List (IUCN 2010) and were based on the IUCN guidelines for regional red lists (IUCN 2003). As with the first one, this new list covers the entire island of Ireland. Both the common frog and smooth newt were considered to be of "Least Concern" while the natterjack toad was assessed as "Endangered". Relevant extracts from this new Red Data List are provided below for each of Ireland's three amphibian species.

Smooth newt (*Lissotriton vulgaris*) - Least Concern

Widespread in Ireland, but locally distributed and under-recorded. May be more common in midlands, but also found in coastal counties (Marnell 1998b; O'Neill et al. *2004). No population estimate available but populations thought to be stable. Although locally distributed, can be abundant where it occurs.*

Common frog (*Rana temporaria*) – Least Concern

Widespread and common throughout Ireland. Found in every county and from sea level to uplands (Marnell 1999). Adaptable species using a broad range of habitats and with a catholic diet (Marnell 1998a). Some evidence of habitat loss (particularly loss of ponds) in Ireland, but no evidence of population decline.

Natterjack toad (*Bufo [Epidalea] calamita*) – Endangered

Previously assessed as Endangered (Whild 1993). In Ireland, restricted to small number of coastal sites on the Dingle and Iveragh peninsulas in western Kerry. While the species' distribution has been relatively stable since the mid-1970s, contraction of its range between 1900 and the 1970s was estimated at 50–60% (Beebee 2002; NPWS 2007). Current range estimated at 76km^2 (NPWS 2007). Also small introduced population in Wexford. Does not occur in Northern Ireland.

Despite recent efforts to improve pond networks around Castlemaine Harbour, the range remains severely fragmented with continuing declines in the quality and extent of habitat in some areas. Estimates of ca. 9,000 breeding adults based on recent intensive three-year monitoring study (Bécart et al. *2007).*

V. Conclusions

Ireland has a limited fauna and only three amphibian species occur there. Both the common frog and the smooth newt are widespread. The third species, the natterjack toad, is restricted to a small number of sites in the southwestern part of the island with a further introduced population in a sand-dune system in the southeast.

The main threats to Irish amphibians concern their habitats, both terrestrial and aquatic. The drainage and infilling of wetlands and the removal of scrub and hedgerows as a result of agricultural improvement and housing developments have resulted in extensive loss and fragmentation of habitats for the newt and frog. Nonetheless, these species are relatively adaptable and have shown an ability to colonize new wetlands such as garden ponds, golf-course ponds, and drainage ditches. Their future in Ireland seems relatively secure.

The listing of the natterjack on Annex IV of the Habitats Directive has provided additional impetus for the conservation of this species. All natterjack breeding sites are protected and while historical losses have begun to be reversed by the recent scheme for creating ponds, further efforts are still required to improve connectivity among sites. Furthermore, ongoing management to ensure the continued suitability of breeding sites and the surrounding terrestrial habitat will remain a challenge.

VI. References

Beebee, T. J. C., 2002. The natterjack toad *Bufo calamita* in Ireland: current status and conservation requirements. *Irish Wildlife Manuals* No. **10**. National Parks and Wildlife Service, Department of the Environment, Heritage and Local Government, Dublin, Ireland.

Bécart, E., Aubry, A. and Emmerson, M., 2007. Monitoring the conservation status of natterjack toad (*Bufo calamita*) in Ireland, 2004 –2006. *Irish Wildlife Manuals* No. **31**. National Parks and Wildlife Service, Department of the Environment, Heritage and Local Government, Dublin, Ireland.

Chrichton, M. 1974. "Provisional distribution maps of amphibians, reptiles and mammals in Ireland". Folens/Foras Forbatha, Dublin.

Dover, W. K., 1877. Toads in Ireland (Natterjack). *Zoologist* **85**: 451.

IUCN, 2003. *Guidelines for Application of IUCN Red List Criteria at Regional Levels:* Version 3.0. IUCN Species Survival Commission. IUCN. Gland, Switzerland and Cambridge, UK.

IUCN, 2010. *Guidelines for using the IUCN Red List Categories and Criteria.* Version 8.1. IUCN. Gland, Switzerland.

King, J. L., Marnell, F., Kingston, N., Rosell, R., Boylan P., Caffrey, J. M., FitzPatrick, Ú., Gargan, P. G., Kelly, F. L., O'Grady, M. F., Poole, R., Roche, W. K. and Cassidy, D., 2011. *Amphibians, Reptiles & Freshwater Fish. Ireland Red List No.* **5**. National Parks and Wildlife Service, Department of Arts, Heritage and the Gaeltacht, Dublin, Ireland.

MacDougald, T. J., 1942. Notes on the habits of the natterjack toad in Co. Kerry. *Irish Naturalists' Journal* **8**: 21–25.

Marnell, F., 1998a. Discriminant analysis of the terrestrial and aquatic habitat determinants of the smooth newt (*Triturus vulgaris*) and the common frog (*Rana temporaria*) in Ireland. *Journal of Zoology* 244: 1–6.

Marnell, F., 1998b. The distribution of the smooth newt (*Triturus vulgaris* L.) in Ireland. *Bulletin of the Irish Biogeographical Society* **22**: 84–96.

Marnell, F., 1999. The distribution of the common frog *Rana temporaria* L. in Ireland. *Bulletin of the Irish Biogeographical Society* **23**: 60 –70.

May, S. and Beebee, T. J. C., 2008. Genetic analysis of Kerry natterjack toad (*Bufo calamita*) populations. Unpublished report prepared for the National Parks and Wildlife Service.

Ni Lamhna, 1979. Provisional Distribution Atlas of Amphibians, Reptiles and Mammals in Ireland. Second Edition. Foras Forbatha, Dublin.

NPWS, 2007. The status of EU protected habitats and species in Ireland – Backing Documents. National Parks and Wildlife Service, Dublin, Ireland. http://www.npws.ie/publications/archive/NPWS_2007_Cons_Ass_Backing_V1.pdf

O'Neill, K., Jennings, S., Forsyth, L., Carey, R., Portig, A., Preston, S. J., Langton T. and McDonald, R., 2004. The distribution and status of smooth newts in Northern Ireland. Unpublished report to Environment and Heritage Service, Northern Ireland.

Whilde, A., 1993. *Threatened mammals, birds, amphibians and fish in Ireland.* Irish Red Data Book **2**: Vertebrates. HMSO, Belfast.

33 Amphibian declines and conservation in Britain

John W. Wilkinson and Richard A. Griffiths

Abbreviations and acronyms used in the text and references:

ARC	Amphibian and Reptile Conservation;
ARG	Amphibian and Reptile Groups
BAP	Biodiversity Action Plan
DEFRA	Department of Environment, Food and Rural Affairs
EEC	European Economic Community
EU	European Union
FCS	Favourable Conservation Status
GIS	Geographical Information System
HMSO	Her Majesty's Stationery Office
IEGB/SPN	Institut d'Ecologie et de Gestion de la Biodiversité Service du Patrimoine Naturel
JARG	Jersey Amphibian and Reptile Group
JDE	Jersey Department of Environment
MPP	Million Ponds Project
NARRS	National Amphibian and Reptile Recording Scheme
NE	Natural England
NERC	Natural Environment and Rural Communities
SAC	Special Areas of Conservation
SAP	Species Action Plan
SSSI	Sites of Special Scientific Interest
UAE	United Arab Emirates
UK	United Kingdom

I. Introduction

This chapter covers the countries of Scotland, England, and Wales, constituting the islands of Britain, i.e., the greater part of the United Kingdom (for Northern Ireland, see Chapter 32, this Issue), and includes notes and information on the British Crown Dependencies of the Isle of Man and the Bailiwicks of Guernsey and Jersey. The total area thus covered is some 230,700 square kilometres. The main island of Great Britain is the third most populous island in the world.

The British climate is temperate oceanic, being generally warmer in winter than adjacent continental Europe due to the moderating effects of the Gulf Stream. Extremes of temperature are rare, however, and the summers are relatively cool. Precipitation can be highly variable, with upland and western areas often receiving the most rain due to North Atlantic depressions; these are most frequent during winter.

Despite the mild climate, the amphibian fauna of Britain is depauperate compared to continental Europe, there being only eight native species (seven in Great Britain and its surrounding islands with an additional species found in Jersey) (Table 33.1). There is, however, a long tradition of herpetological study in Britain, with amphibian declines being noted from the middle of the 20th century (e.g. Cooke 1972; Beebee 1975, 1976), prior to the "global phenomenon" of amphibian declines that have deservedly received much recent attention (see for example Beebee and Griffiths [2000] and chapters in Heatwole and Wilkinson [2009]). Early amphibian declines in Britain can be primarily attributed to changes in land use, including development and agricultural intensification following the Second World War (*sensu* Carrier and Beebee 2003). These factors continue to represent challenges for the conservation of amphibians in Britain and are magnified by additive threats such as climatic change and emerging infectious diseases. Great Britain is a relatively small and intensively developed country that lacks the extensive depopulated protected areas characteristic of some continents. Consequently, conservation relies on maintaining or creating habitats for amphibians within landscapes that are primarily agricultural or otherwise developed. This chapter treats those British amphibians for which declines have been of greatest cause for concern,

Table 33.1 Status of native British amphibians.

P = present   **PP** = present/protected   ↔ = trend stable   ↑ = trend up   ↓ = trend down   ? = trend unknown

SPECIES	SCOTLAND	WALES	ENGLAND	ISLE OF MAN	BAILIWICK OF GUERNSEY	BAILIWICK OF JERSEY
Rana temporaria	P↔	P↔	P↔	PP?	P↔	-
Rana dalmatina	-	-	-	-	-	PP↑
*Pelophylax lessonae**	-	-	PP?	-	-	-
Bufo bufo	P?	P↓	P↓	-	-	PP↓
Bufo calamita	PP?	PP↑	PP↔**	-	-	-
Triturus cristatus	PP↓	PP↓	PP↓	-	-	-
Lissotriton vulgaris	P?	P?	P?	-	P↔#	-
Lissotriton helveticus	P?	P?	P?	-	P?##	PP?$

*	Native only in England; populations of non-native origin also present in England and elsewhere in Britain
**	Populations down at some English sites but increasing at others
#	Declining in some areas of Guernsey but expanding in range
##	Present on the island of Alderney
$	Probably declining

and describes the conservation and monitoring measures implemented in attempts to address those concerns.

II. Amphibian declines and conservation measures

A. The great crested newt (*Triturus cristatus*)

A complex array of legal instruments enshrines the great crested newt in UK law. In summary, the species is afforded strict protection in Britain through a combination of the Wildlife and Countryside Act 1981 (as amended) and the Conservation of Habitats and Species Regulations 2010. It is also listed as a species of principal importance for the conservation of biodiversity in both England and Wales under Sections 41 and 42 (respectively) of the Natural Environment and Rural Communities (NERC) Act 2006 – a Biodiversity Action Plan (BAP) priority species. These legal instruments afford protection for individual newts, populations, and habitats, and an offence may be committed if disturbance of newts occurs at any of these levels. In addition, site-specific protection is afforded through designations of important areas as Sites of Special Scientific Interest (SSSIs) or Special Areas of Conservation (SACs). Although it is estimated that less than 1% of the UK population of great crested newts occurs in such areas (Jehle *et al.* 2011), both SSSIs and SACs provide a relatively high degree of legal protection against disturbance or habitat loss.

The Conservation of Habitats and Species Regulations 2010 implements the European Union's 'Habitats Directive' (Council Directive 92/43/EEC (a) on the Conservation of Natural Habitats and of Wild Fauna and Flora) in Great Britain; the great crested newt is listed on Annexes II and IV of the Directive. As a consequence of this strict protection, the United Kingdom has legal obligations to report on the conservation status of the great crested newt every six years, with the specific reporting requirements set out in Article 17. Moreover, article 11 of the Habitats Directive states that good knowledge of a species (e.g., range/distribution, occurrence, biology, ecology, threats and sensitivity, conservation needs) and regular surveillance of its conservation status over time are essential preconditions for any meaningful conservation strategy.

This strict protection for a (still) widespread species has arisen in Britain because of the relative importance of British populations in the European context and because of the considerable declines the species has suffered over the past century (Langton *et al.* 2001). The species appears especially vulnerable to the usual agents of amphibian decline, i.e., loss and alteration of habitat (e.g. Gasc *et al.* 1997). In Britain, agricultural intensification and development, especially in great crested newt "hotspots" such as northwestern England, northeastern Wales, and parts of southern Britain, have added to the landscape-scale loss and degradation of ponds to contribute to the declines of this species. Indeed, some 15 years before global amphibian declines became a conservation issue, Beebee (1975) reported a 50% loss of populations between 1960 and 1975 and pointed out that local extirpations were already occurring. Various studies since then have confirmed continued loss of habitat (Jehle *et al.* 2011), with the annual loss of breeding ponds estimated to be up to 2% (UK Biodiversity Steering Group 1995). Ongoing loss of habitat, coupled with greater awareness of the issues, means that great crested newts are coming into conflict with development more frequently than ever before. This means that issues of conservation and planning pertaining to this species and its habitats are often prominent, occasionally controversial, and may have far-reaching financial implications.

Mitigation projects, designed to minimize the impact of development and create aquatic and terrestrial habitats for great crested newts in compensation for areas affected, are now frequently employed in an attempt to maintain the status of the species in Britain. Unfortunately, however, these projects do not always achieve this objective and, most pertinently, many lack funding (and/

or the commitment) for the long-term, ongoing monitoring required to assess their success (or otherwise). What is clear is that populations of great crested newts have often ended up in small, fragmented habitats surrounded by development. An implicit assumption of such mitigation projects is that a great crested newt population can be maintained – and possibly enhanced – in smaller patches of habitat if these areas are improved for the species. Although this may be possible in the short-term to medium-term, the long-term viability of populations that are isolated and close to dispersal barriers is uncertain (Lewis *et al.* 2007). In 2001, revised guidance for developers was produced in an effort to improve the methods, assessment, and outcomes of mitigation projects (English Nature 2001). A three-year project, funded by the UK Department of Environment, Food and Rural Affairs (DEFRA) and carried out by Amphibian and Reptile Conservation (ARC) and the University of Kent has just been initiated (2011) to investigate whether projects that have been carried out under the new guidance have resulted in more favourable conservation outcomes for the species.

Other conservation measures include pond-creation schemes, such as the Million Ponds Project (MPP), a collaborative project organized by Pond Conservation, but with a large number of organizations and land managers as partners to this challenging task. ARC is a key partner, coordinating the herpetological element, with the remit to create ponds suitable for amphibians and reptiles. Due to the importance of maintaining the existing pond stock, particularly for great crested newts, management of ponds is also included.

Sustainable management of the farmed environment is essential to landscape-level conservation projects, and may be achieved best through agri-environmental schemes. To maximize the conservational benefit of land management/creation options within the various schemes, ARC has produced leaflets on agri-environments to highlight the best options for great crested newts (and reptiles) and the key principles to be considered, including such topics as where to locate new features.

Assessment of the conservation status of *T. cristatus* in Scotland, Wales and England is currently being improved through several ARC projects in partnership with the UK's Statutory Bodies. These use GIS-based modelling techniques to predict species' presence and the suitability of habitat using area-specific models. This is particularly useful in areas for which survey information is incomplete or out-of-date (sometimes as a result of low availability of data due to the monetary value or confidentiality of data, which may have originated from mitigation projects!). New BAP conservation targets should ultimately result from this more accurate information, allowing more cost-effective, regionally and locally targeted action. Modelling has already demonstrated that Scottish *T. cristatus* populations are likely very uncommon and the species in Scotland may be more completely known than was previously thought (see Wilkinson *et al.* 2011). The most recent efforts describe ways of assessing local Favourable Conservation Status (FCS) within an administrative area or region, in this case Flintshire in northeastern Wales, and suggesting means whereby targeted creation of ponds and habitats can contribute to restoration of FCS for the species (Arnell and Wilkinson, 2011).

B. The pool frog (*Pelophylax lessonae*)

The pool frog is (now) protected under the same legislation as the great crested newt. There have been documented introductions of pool frogs (and other *Pelophylax* spp.) to Britain since the 19th Century (e.g. see Smith 1951) but, by the early years of the 21st Century, considerable evidence had accumulated that some populations of the species in eastern England were/had been of native origin (Beebee *et al.* 2005). This followed research arising from the establishment of a pool frog Species Action Plan and Steering Group (SAP) (English Nature 1998). It was determined, moreover, that English pool frogs formed part of a "northern clade" of the species, being most closely related to those pool frogs found in Norway and Sweden. Unfortunately, during the short time over which this research occurred, the last native population known in England had declined to extirpation (Buckley and Foster 2005).

Declines of northern pool frog populations in England are thought to have begun with large-scale drainage of the East Anglian fenlands from the 17th century onwards and, latterly, resulted from dry summers and water abstraction and changes in habitat management leading to the loss of the pingo ponds on which the species depended (Buckley and Foster 2005). A Pool Frog SAP Steering Group was established by Natural England (NE), ARC and Anglian Water in order to effect the reintroduction of northern-clade pool frogs to suitably-restored and managed sites in East Anglia, using donor populations in Sweden. These were the most robust of the potential northern-clade donor populations (Buckley and Foster 2005). Following considerable research on habitat assessment and an analysis of population viability, reintroductions were carried out at an appropriate Norfolk site for four years, beginning from 2005. A ten-year habitat management agreement is in place at this site and further potential locations for reintroductions are being evaluated within the species' historical range.

C. The natterjack toad (*Bufo [Epidalea] calamita*)

This species also enjoys full protection in the UK and although it receives the same level of protection as the great crested newt, it comes into conflict with development much less frequently as a result of its restricted distribution and specialist habitat requirements. Natterjacks have declined, however, through loss of habitat (e.g. dune, upper saltmarsh, and heathland) and changes in traditional land use, including building, afforestation, drainage of land, and the disruption of land-forming coastal processes (Beebee 1977). Losses of populations on heathland have been much greater that on dune or upper saltmarsh due, at least in part, to the contemporary low connectivity of many such sites. The natterjack is a pioneering species that has evolved in ways allowing it to cope with local loss of populations through re-colonization of breeding sites, according to the dynamics of local conditions (e.g. loss and formation of new dune slacks). Coastal sites more often have the inherent connectivity required by this species' metapopulations than do heathlands, which are often found in urbanized landscapes or otherwise are surrounded by inimical habitat.

All established breeding sites, numbering about 69 in total, are considered to be known and breeding success is monitored at each of these every year, so far as is possible (Buckley and Beebee 2004). In total, translocations have been carried out at 29 sites since 1975 and, of these, 27 are at a stage at which the level of success can be judged. Nineteen of the 27 (70%) have been successful at least in the short to medium term (Griffiths *et al.* 2010). Re-establishing natterjacks on heathland (57% success) has proved much more difficult than on dunes and upper saltmarshes (85% success) because heathland tends to require greater input for habitat management. Details of the habitat requirements of natterjacks in Britain and the methodology for translocations are given by Baker *et al.* (2011).

Natterjacks at some sites have been shown to be infected with the chytrid fungus *Batrachochytrium dendrobatidis* but it is not yet known whether this pathogen is responsible for any observed population declines (e.g. on the Solway Firth in Scotland). A project to monitor the spread and effects of this chytrid in Britain, coordinated by the Institute of Zoology, is ongoing (see also Garner, this issue, and http://www.zsl.org/conservation/regions/uk-europe/ukchytridiomycosis,842,AR.html).

D. The common toad (*Bufo bufo*)

Common toads are protected only against sale in the UK (i.e., not against killing or habitat destruction) but are a Biodiversity Action Plan (BAP) priority species because of recent recorded declines, especially in central and southern England (Carrier and Beebee 2003). Under the NERC Act, developers have to consider the effects of any development on BAP priority species, but this is often all they do! Most developments that affect common toad populations in the UK tend to go ahead with either loss of the population, or little or no mitigation.

Loss of aquatic and/or terrestrial habitat in areas under high development pressure is probably responsible for many declines but decreasing connectivity between suitable habitat patches is also important. Hitchings and Beebee (1998) have shown that this species fares poorly in urbanized landscapes in comparison to rural breeding sites and in comparison to other species with super-ficially similar ecology, such as the common frog (Brede and Beebee 2004; 2006). The species has also declined for similar reasons in Jersey (Wilkinson *et al.* 2007) where it was once abundant. Common toads now persist there mainly in quite small populations that have become isolated by development. They appear to have adapted to utilizing small, garden ponds for reproduction but the resulting population structure is vulnerable to local extirpations. Further research into declines of common toads, especially in lowland England, is required.

E. The agile frog (*Rana dalmatina*)

This species occurs in Britain only on the island of Jersey where, like all herpetofauna there, it enjoys full protection under the Conservation of Wildlife (Jersey) Law 2000 (as amended). Agile frogs once occurred at several locations, mainly in the west of the island (Le Sueur 1976) but, by the early 1990s, this had been reduced to one remaining breeding site at Ouaisné Common.

A collaborative effort between Jersey Department of Environment (JDE) and Durrell (Jersey Zoo) and with assistance from Jersey Amphibian and Reptile Group and habitat management by ARC staff, has since begun to reverse the fortunes of the species in Jersey. A combination of *in situ* protection of spawn and head-starting of tadpoles in a biosecure unit at Durrell appears to have been successful. Tadpoles were retained until metamorphosis and released at existing, re-profiled and newly created ponds at Ouaisné as well as at one other former site. Both breeding sites now appear to be faring well, the number of spawn clumps at Ouaisné being up by about 500% com-pared to five years ago (JARG 2010), and more breeding ponds on the site are now occupied. The main issue for the agile frog on Jersey lies with trying to expand the range of the species on a small, highly developed island with a shortage of suitable habitat.

F. Other species and issues

Of the remaining native British species (*Rana temporaria, Lissotriton vulgaris,* and *L. helveticus*), there is protection only against sale, and none are BAP priorities. An EU reporting requirement exists for *R. temporaria* because of the species' exploitation for food within the EU, although this attracts no extra protection in the UK. Breeding sites of these species (and of common toads) are thus available to be developed!

Although the chytrid fungus has not yet been blamed for any large-scale declines in Britain, research is ongoing (see above). The last round of surveying of chytrids in the UK (Cunningham and Minting 2009) found them at sites across Britain and a strong association between infection at a site and the presence of non-native amphibians. Such introduced alien species (particularly *Lithobates catesbeianus, Ichthyosaura alpestris, Xenopus laevis,* and possibly *Pelophylax ridibundus*) may have been routes of infection for the disease. Species such as *I. alpestris,* now present at about 30 locations in mainland Britain, also appear to be asymptomatic vectors. The pattern for *I. alpestris* appears to be one of establishment at the point of introduction and low-level dispersal into other pools within the area, but not to the extent that it displaces native species of newts (Bell and Bell 1993; Suzuki 2009). *Lithobates catesbianus* and *X. laevis* are considered to have the greatest potential impact on native species, both as possible vectors of disease (Garner *et al.* 2006) and as predators/

competitors (see e.g. Measey and Tinsley 1998). These species have been the subject of attempts at eradication in England and Wales, as well as in other parts of Europe. At least some of these attempts have met with some success (Ficetola *et al*. 2007).

The effects of other exotics currently established in the UK are largely unknown, the exception being *Triturus carnifex*, which has hybridized with native *T. cristatus* at at least one of its two known establishment sites. Pure or hybrid *T. carnifex* are, however, not thought to have spread far from the original site of introduction (Brede *et al*. 2000). There are similar concerns about any *T. marmoratus* that may be (or will become) established in Britain.

About ten populations of *Alytes obstetricans,* resulting from documented introductions over a century ago, are known from the southern half of Britain but are not currently known to have an impact on native species, although they do give rise to complaints about noise (usually in urban areas) where they turn up! The recording of exotic amphibians in Britain is discussed below.

G. Species' status
There is no Red List for British amphibians. Amphibians native to Britain and their status in the various jurisdictions are given in Table 33.1.

III. Monitoring of amphibians in Britain
Herpetological recording in Britain has a long history, with the first set of distribution maps being published over 60 years ago (Taylor 1948). Updated maps and accounts of distributions have been published at intervals since then (e.g. Taylor 1963; Arnold 1973, 1983, 1996; Swan and Oldham 1993). The early accounts were more collations of *ad hoc* records than reports of co-ordinated monitoring programmes. National programmes to collect and analyze monitoring data in a standardized way are therefore a more recent development, but have suffered from fixed-term funding and shifting priorities within the governmental agencies responsible for such projects. Nevertheless, there is now a well-established network of voluntary Amphibian and Reptile Groups (ARGs) in Britain. These groups usually cover individual counties and many are heavily engaged in the collection, analysis, and dissemination of regional amphibian-monitoring data.

At the national level, monitoring of amphibians (and reptiles) in Britain is currently carried out under the umbrella of the National Amphibian and Reptile Recording Scheme (NARRS). This brings together pre-existing monitoring efforts with attempts to gain a handle on status, and changes in status, of the widespread species. *Bufo calamita* and *P. lessonae* are monitored annually at every breeding pond, with spawn production and the success of metamorphosis being recorded. These efforts are coordinated by ARC. JDE operates a similar monitoring programme for *R. dalmatina*.

NARRS Widespread Amphibian Surveys began in 2007, the aim being to generate sufficient survey data by 2012 to provide a "status baseline" based on species' occupancy of ponds in randomly-generated 1-km grid squares. For full protocols, see www.narrs.org.uk and Wilkinson and Arnell (2011). The latter also summarize results from the surveys of 2007–2009. NARRS results for Jersey to date are given by Wilkinson and Arnell (2010). Emphasis in NARRS Widespread Species Surveys is on adherence to robust, repeatable protocols so that data from future survey cycles are directly comparable and inferences regarding change in status can be made. Results are also broken down regionally. Objective assessments of habitat quality are also carried out during each survey, based on the Habitat Suitability Index developed for the great crested newt (Oldham *et al*. 2000). As the scheme progresses, it is intended that any perceived changes in status can be related to recorded differences in habitat quality and that the data can ultimately be used to guide conservation policy and action.

Alien amphibians are also recorded (when encountered) during NARRS surveys, and a dedicated website has been established to allow casual reports of alien herpetofauna to be submitted (www.alienencounters.org.uk).

IV. Conclusions

Throughout Britain, by far the greatest threat to amphibian populations remains the loss and fragmentation of habitats. This is particularly true in Britain's designated "growth areas" where amphibian populations that may have been established for many decades on "under-utilized", brownfield and agricultural land, are especially threatened by new housing and the development of infrastructure. Although policy and legislation theoretically protect the most vulnerable species, development is, in practice, still negatively impacting populations of species such as *T. cristatus*. Ongoing research into mitigation efforts, assessment of status, and monitoring are intended to improve this situation in the long term.

In order to protect the relatively common, widespread, and less-protected British amphibians, legislation in the UK against killing (as is already extended to all British reptiles) would be beneficial, coupled with the establishment of more reserves for exceptional populations of species like the common toad. Further research into declines of toads and into potential long-term effects of emerging infectious diseases are also still required. Synergistic interactions between the known, existing threats and imminent potential problems, such as climatic change, also need anticipation and consideration. The persistence of robust amphibian populations in Britain ultimately depends on such targeted research, repeatable monitoring, and directed conservation actions in a better-connected landscape.

V. Acknowledgements

We would like to thank John Buckley, Dorothy Wright, and Andy Arnell (all of ARC), Trevor Beebee (University of Sussex), and Jane Gilmour (Guernsey Biological Records Centre) for valuable contributory information and their comments on parts of the manuscript. Assessment and modelling of the status of great crested newts is supported by the Countryside Council for Wales, Natural England and Scottish Natural Heritage. NARRS Reports were supported by Amphibian and Reptile Groups of the UK and Jersey Department of Environment. For a full list of NARRS partners, see www.narrs.org.uk.

VI. References

Arnell, A. P. and Wilkinson, J. W., 2011. *Pilot Modelling to Inform Determination of Favourable Conservation Status for the Great Crested Newt.* Countryside Council for Wales Contract Science Report No. 961.

Arnold, H. R., 1973. *Provisional Atlas of the Amphibians and Reptiles of the British Isles.* Biological Records Centre, Abbots Ripton, UK.

Arnold, H. R., 1983. *Distribution Maps of the Amphibians and Reptiles of the British Isles.* Biological Records Centre, Abbots Ripton, UK.

Arnold, H. R., 1996. *Atlas of Amphibians and Reptiles in Britain.* ITE Research Publication 10. HMSO, London.

Baker, J., Beebee, T., Buckley, J., Gent, T. and Orchard, D., 2011. *Amphibian Habitat Management Handboo".* ARC, Bournemouth.

Beebee, T. J. C., 1975. Changes in the status of the great crested newt *Triturus cristatus* in the British Isles. *British Journal of Herpetology* **5**: 481–490.

Beebee, T. J. C., 1976. The natterjack toad (*Bufo calamita*) in the British Isles; a study of past and present status. *British Journal of Herpetology* **5**: 515–521.

Beebee, T. J. C., 1977. Environmental change as a cause of natterjack toad *Bufo calamita* declines in Britain. *Biological Conservation* **11**: 87–102.

Beebee, T. J. C. and Griffiths, R. A., 2000. *Amphibians and Reptiles.* HarperCollins, London.

Beebee, T. J. C., Buckley, J., Evans, I., Foster, J., Gent, A. H., Gleed-Owen, C. P., Kelly, G., Rowe, G., Snell, C., Wycherley, J. T. and Zeisset, I., 2005. Neglected native or undesirable alien. Resolution of a conservation dilemma concerning the pool frog Rana lessonae. *Biodiversity and Conservation* **14**: 1607–1626.

Bell, A. P. and Bell, B. D., 1993. Distribution of the introduced alpine newt *Triturus alpestris* in Shropshire, England, and its impact on the native *T. cristatus* and *T. vulgaris*. P. 21 in *Abstracts of the Second World Congress of Herpetology*. Adelaide.

Brede, E. G. and Beebee, T. J. C., 2004. Contrasting population structures in two sympatric anurans: implications for species conservation. *Heredity* **92**: 110–117.

Brede, E. G. and Beebee, T. J. C., 2006. Consistently different levels of genetic variation across the European ranges of two anurans, *Bufo bufo* and *Rana temporaria. Herpetological Journal* **16**: 265–271.

Brede, E. G., Thorpe, R. S., Arntzen, J. W. and Langton, T. E. S., 2000. A morphometric study of a hybrid newt population (*Triturus cristatus/T. carnifex*): Beam Brook Nurseries, Surrey, U.K. *Biological Journal of the Linnean Society* **70**: 685–695.

Buckley, J. and Beebee, T. J. C., 2004. Monitoring the conservation status of an endangered amphibian: the natterjack toad *Bufo calamita* in Britain. *Animal Conservation* **7**: 221–228.

Buckley, J. and Foster, J., 2005. *Re-introduction Strategy for the Pool Frog Rana lessonae in England.* English Nature Research Report no. **642**, English Nature, Peterborough.

Carrier, J.-A. and Beebee, T. J. C., 2003. Recent, substantial, and unexplained declines of the common toad *Bufo bufo* in lowland England. *Biological Conservation* **111**: 395–399.

Cooke, A. S., 1972. Indications of recent changes in status in the British Isles of the frog (*Rana temporaraia*) and the toad (*Bufo bufo*). *Journal of Zoology* **167**: 161–178.

Cunningham, A. A. and Minting, P., 2009. *National Survey of Batrachochytrium dendrobatidis Infection in Amphibians, 2008.* Institute of Zoology, Zoological Society of London, London.

English Nature, 1998. *Pool Frog Rana lessonae Species Action Plan.* English Nature, Peterborough, UK.

English Nature, 2001. *Great Crested Newt Mitigation Guidelines.* English Nature, Peterborough, UK.

Ficetola, G. F., Coïc, C., Detaint, M., Berroneau, M., Lorvelec, O. and Miaud, C., 2007. Pattern of distribution of the American bullfrog *Rana catesbeiana* in Europe. *Biological Invasions* **9**: 767–762.

Garner, T. W. J., Perkins, M. W., Govindarajulu, P., Seglie, D., Walker, S., Cunningham, A. A. and Fisher, M. C., 2006. The emerging amphibian pathogen *Batrachochytrium dendrobatidis* globally infects introduced populations of the North American bullfrog, *Rana catesebiana. Biology Letters* 2006 **2**: 455–459.

Gasc, J.-P., Cabela, A., Crnobrnja-Isailovic, J., Dolmen, D., Grossenbacher, K., Haffner, P., Lescure, L., Martens, H., Martinez Rica, J. P., Maurin, H., Oliveira, M. E., Sofianidou, T. S., Veith, M. and Zuiderwijk, A. (eds), 1997. *Atlas of Amphibians and Reptiles in Europe.* Societas Europaea Herpetologica & Museum National d'Histoire Naturelle (IEGB/SPN), Paris.

Griffiths, R. A., McGrath, A. and Buckley, J., 2010. Re-introduction of the natterjack toad in the UK. Pp. 62 – 65 in *Global Re-Introduction Perspectives: Additional Case-studies around the Globe,* ed by P. S. Soorae. IUCN/SCC Re-introduction Specialist Group, Abu Dhabi, UAE.

Heatwole, H. and Wilkinson, J. W. (eds), 2009. *Amphibian Decline: Diseases, Parasites, Maladies and Pollution.* Vol. 8 in the series *Amphibian Biology,* ed by H. Heatwole. Surrey Beatty and Sons, Baulkham Hills.

Hitchings, S. P. and Beebee, T. J. C., 1998. Loss of genetic diversity and fitness in common toad (*Bufo bufo*) populations isolated by inimical habitat. *Journal of Evolutionary Biology* **11**: 269–283.

JARG, 2010. Agile frog taddy boom! *Jargon: Newsletter of the Jersey Amphibian and Reptile Group* **Autumn 2010**: 1.

Jehle, R., Thiesmeier, B. and Foster, J., 2011. *The Crested Newt. A Dwindling Pond-Dweller.* Laurenti-Verlag, Bielefeld, Germany.

Langton, T., Beckett, C. and Foster, J., 2001. *Great Crested Newt Conservation Handbook.* Froglife, Halesworth UK.

Le Sueur, F., 1976. *A Natural History of Jersey.* Phillimore and Co., London.

Lewis, B., Griffiths, R. A. and Barrios, Y., 2007. *Field Assessment of Great Crested Newt* Triturus cristatus *Mitigation Projects in England.* Natural England Research Report **NER001**, Natural England, Sheffield, UK.

Measey, G. J. and Tinsley, R. C., 1998. Feral *Xenopus laevis* in South Wales. *Herpetological Journal* **8**: 23–27.

Oldham, R. S., Keeble, J., Swan, M. J. S. and Jeffcote, M., 2000. Evaluating the suitability of habitat for the great crested newt (*Triturus cristatus*). *Herpetological Journal* **10**: 143–155.

Smith, M., 1951. *The British Amphibians and Reptiles.* Collins, London.

Suzuki, M., 2009. *Dispersal of an Alien Amphibian* (Mesotriton alpestris*) from its Introduction Site.* MSc dissertation, University of Kent.

Swan, M. J. S. and Oldham, R. S., 1993. *Herptile Sites Volume 1: National Amphibian Survey Final Report.* English Nature Research Reports **38**, English Nature, Peterborough, UK.

Taylor, R .H. R., 1948. The distribution of amphibians and reptiles in the British Isles, with notes on species introduced. *British Journal of Herpetology* **1**: 1–38.

Taylor, R. H. R., 1963. The distribution of amphibians and reptiles in England and Wales, Scotland and Ireland and the Channel Islands: a revised survey. *British Journal of Herpetology* **3**: 95–115.

UK Biodiversity Steering Group, 1995. *Biodiversity: The UK Steering Group Report.* Volume 2: *Action Plans.* HMSO, London.

Wilkinson, J. W. and Arnell, A. P., 2010. *Jersey NARRS Report 2010: Results of the National Amphibian and Reptile Recording Scheme in Jersey, 2007–2010.* Unpublished Report to the States of Jersey Planning and Environment Department.

Wilkinson, J. W. and Arnell, A. P., 2011. *NARRS Report 2007 – 2009: Interim results of the UK National Amphibian and Reptile Recording Scheme Widespread Species Surveys.* ARC Research Report **11/01**.

Wilkinson, J. W., Beebee, T. J. C. and Griffiths, R. A., 2007. Conservation genetics of an island toad: *Bufo bufo* in Jersey. *Herpetological Journal* **17**: 192–198.

Wilkinson, J. W., Wright, D., Arnell, A. and Driver, B., 2011. *Assessing population status of the great crested newt in Great Britain.* Natural England Commissioned Reports, Number **080**, Natural England, Sheffield, UK.

34 Conservation and declines of amphibians in The Netherlands

Anton H. P. Stumpel

Abbreviations and acronyms used in the text and references:

CBS	*Centraal Bureau voor de Statistiek (Statistics Netherlands)*
EHS	*Ecologische Hoofdstructuur (National Ecological Network)*
IUCN	*International Union for the Conservation of Nature*
LNV	*Former name of the Netherlands' Ministry of Economic Affairs, Agriculture and Innovation*
NGO	*Non-governmental organization*
RAVON (foundation)	*Reptile, Amphibian & Fish Conservation Netherlands*
TRIM	*Trends and Indices for Monitoring Data.*

I. Introduction

The Netherlands is a small, densely populated country that lies in northwestern Europe at approximately 52° N and 5° E. It is bordered by the North Sea to the north and west, Germany to the east, and Belgium to the south. With 16 million inhabitants on a land area of 33,874 km² (population density of 472/km²), the landscape is largely man-made, intensively cultivated, and urbanized. The western and northern parts are a river delta where the estuaries of the Rhine, Meuse, and Scheldt come together. Water is thus an important feature of the landscape. Reclamation of land began in the 13th century when man radically changed the landscape; dykes were built, and polders with a dense network of ditches and canals were created. Apart from some marshes, no primary or pristine habitats were left. Where there was no natural freshwater on the surface, ponds were dug. Much land is low-lying, nearly a quarter of the total surface area being below sea level, the lowest at -7 m. The soil consists of various substrates, especially clay, sand, peat, marl, and gravel; there are no rocks. Most of the country is quite flat, the highest point being only 321 metres above sea level. The Netherlands has an Atlantic climate with an annual precipitation of 780 mm.

Nowadays, most of the land surface has been drained for agriculture, construction of houses, industry, and for supply of drinking water. The landscape is drastically fragmented by many roads, canals, and railways. Both water and soil are strongly eutrophic due to airborne pollution.

The Netherlands has sixteen indigenous amphibian species and some of them reach the limits of their European range there. *Salamandra salamandra*, *Alytes obstetricans*, and *Bombina variegata* reach their northern limit in The Netherlands and are confined to the southeasternmost part of the country. *Ichthyosaura alpestris*, *Hyla arborea*, *Pelobates fuscus*, and *Rana* kl. *esculenta* reach their northwestern European extent. *Rana lessonae* does likewise, apart from a single isolated population in England. The Netherlands' North Sea coast forms the western limit of the distribution area of *Rana arvalis*. Additionally, in The Netherlands *Lissotriton helveticus* is distributed only in the southern provinces, although this newt has a more northern distribution in Britain and Germany.

II. Declining species and species of special conservation concern

In The Netherlands, all indigenous amphibians have been legally protected under the Nature Conservation Act since 1973. As well as naming species, this act also designated some sites as *Beschermde Natuurmonumenten* (Protected Nature Monuments). The Flora and Fauna Act (Backes and Verschuuren 2001) provided an update of the Nature Conservation Act and included the provisions of the Bern Convention and the Habitats and Birds Directives. Nonetheless, the new Act has exemptions that provide loopholes, leading to a political tug-of-war when people apply for planning permission. This can lead to important habitats being destroyed. Recently, a new national Red List for reptiles and amphibians was published (van Delft *et al.* 2007).

Amphibians in The Netherlands are seriously under threat from the results of human activity and also from disease. Amphibians' habitats, both terrestrial and aquatic, are being affected by large-scale land use, intensive animal husbandry, drainage, mechanical and frequent mowing and cutting, frequent crop rotation, overgrazing, and the use of fertilizers, herbicides, and pesticides (Stumpel 2004). Fragmentation and loss of habitat is also the result of the ever-increasing urbanization, with a landscape dissected by roads, canals, and railways. Moreover, there is little room for well-developed habitats for amphibians because of the practices of modern water management, whereby watercourses are straightened and frequent dredging occurs, and banks are mown. Ironically, also in protected areas, mismanagement sometimes happens. Lack of money, time, and skills lead to lack of management and to mechanical cutting, bad timing of measures, too frequent intervention, and overgrazing. In more general terms, amphibian decline can also be attributed to predation by pets, waterfowl, or fish that have been released in ponds or ditches, as well as to the pollution of waters used for breeding. Collectors also take their toll of the wild population (cf. Stumpel, 2004). Infectious diseases form a threat of a different order. People have only recently become aware of the present and potential threat of chytridiomycosis and the disease caused by *Ranavirus* (Spitzen-van der Sluijs *et al.* 2010). Much research is still needed for developing a conservation policy that takes these threats into account (cf. Garner, this Volume).

All these factors have led to half of the indigenous species being on the national Red List (van Delft *et al.*, 2007); their present status according to both Netherlands' criteria (trend since 1950, rareness) and those from IUCN, is briefly indicated in Table 34.1. For more information, see the "Atlas of the Herpetofauna of The Netherlands" (Creemers and van Delft 2009).

Table 34.1 The status of amphibians in The Netherlands, assessed at the national level (STATUS-NL) and according to IUCN criteria (STATUS-IUCN) (van Delft *et al.* 2007).

SPECIES	STATUS - NL	STATUS - IUCN
Bombina variegata	Critically endangered	Endangered
Salamandra salamandra	Endangered	Endangered
Hyla arborea	Endangered	Least concern
Pelobates fuscus	Endangered	Endangered
Triturus cristatus	Vulnerable	Least concern
Lissotriton helveticus	Vulnerable	Vulnerable
Alytes obstetricans	Vulnerable	Vulnerable
Bufo calamita	Susceptible	Vulnerable

A. *Bombina variegata*

The yellow-bellied toad (*Bombina variegata*) is confined to the southernmost part of the country. Only five populations, now inhabiting abandoned quarries, have survived the destruction and fragmentation of the species' habitat. Recent introductions into newly developed habitats in those quarries have been successful. However, as long as the total population numbers less than 500 adult individuals, the long-term survival of the species remains uncertain. Population densities may fluctuate strongly between years. Appropriate habitat management for this toad includes the repeated creation of shallow, temporary waters shortly before their reproductive season.

B. *Salamandra salamandra*

The fire salamander (*Salamandra salamandra*) is restricted to the same area as *B. variegata*. It is a terrestrial ovoviviparous species of deciduous forests that contain streams and springs. The free-swimming larvae need clear, oxygen-rich water that is free of fish, and there must be plants, stones, or logs to serve as shelter. Threats originate from humans spoiling the habitat by clearing away aquatic plants, polluting the water, and straightening streams or making sluices in them. In addition, some animals are caught by collectors and still others from foreign populations are set free, thus releasing alien genetic material into the native population.

C. *Hyla arborea*

The European treefrog (*Hyla arborea*) is found in the southeastern half of the country. Its terrestrial habitat requires subtle transitions from one vegetation type to another, such that the quality of the terrestrial habitat determines whether a species is present in an area more than does suitable aquatic habitat. This species reproduces mainly in marshes and ponds. In the cultural landscape, habitats for reproduction, aestivation, and hibernation may be hundreds of metres apart so good connectivity among them is vital for survival. Also important is the position of a single habitat in the network of habitats of the metapopulation. Population densities may fluctuate strongly between years. Although *H. arborea* has been successfully introduced into newly developed habitats during the past 35 years, the species is still suffering elsewhere from the effects of drainage, habitat fragmentation and from being collected. Moreover, the aquatic habitat remains under threat from infilling, lack of management, pollution, and release of fish.

D. *Pelobates fuscus*

The European spadefoot toad (*Pelobates fuscus*) is close to becoming classified as critically endangered. Being a burrowing amphibian, many of its terrestrial habitats in loose sandy soils have become unsuitable due to changed use of agricultural land. Apart from isolation and loss of habitat, numbers are declining and populations are rapidly disappearing. However, a few small populations survive in the east of the country. Because spadefoots reproduce in old ponds and pools, they are susceptible to neglect of management programmes, including the removal of released fish, especially pumpkinseed sunfish (*Lepomis gibbosus*). Removal of plants both in and around the pool for management of the flora can also have a disastrous effect. Unfortunately, the species' low rating in international conservation legislation has led to little interest from the government and, consequently, appropriate management does not take place.

E. *Triturus cristatus*

The aquatic habitat of the great crested newt (*Triturus cristatus*) is similar to that of *Hyla arborea* and therefore the species is susceptible to the same threats. Their terrestrial habitat in such places as rough grassland, gardens, scrub, and deciduous woodland, need to be close to the aquatic sites used for reproduction. Consequently, the species is threatened by its aquatic and terrestrial habitats being isolated from each other, for example by roads. This leads to many newts becoming traffic casualties during their migrations in the early spring and in the autumn, and probably has an impact on local populations. The creation of new ponds near potential terrestrial habitat has been shown to be successful for this species.

F. *Lissotriton helveticus*

The palmate newt (*Lissotriton helveticus*) is limited to the southern part of The Netherlands, where it is mostly found in natural areas. This species is tolerant of a low pH in the heathland, moorland, and woodland pools where it reproduces. The drastic decline of this newt during the past 50 years can be attributed to large-scale reclamation of heathland and moorland for agriculture, forestry, and housing; the remaining habitats have become fragmented and drained. Air pollution causes the water in which it reproduces to become acidified to a pH below tolerance levels and has further contributed to its decline.

G. *Alytes obstetricans*

The midwife toad (*Alytes obstetricans*) has the same, limited distribution as *Bombina variegata*. It is a rare species in The Netherlands. It requires a loose, stony soil in a warm location. These conditions can be found on agricultural land that is traditionally managed, in quarries, and in old churchyards. A strong decline has recently been observed that cannot be linked to changes in habitat; perhaps the cause can be attributed to chytrid infection, as is described for Spain (Bosch *et al.* 2001).

H. *Bufo calamita*

The natterjack toad (*Bufo calamita*) occurs in various types of habitat such as heathlands, coastal sand dunes, edges of salt marshes, marshes in river forelands, and quarries, as well as in temporary habitats left by man after excavating ground for building (Stumpel, 2004). Apart from heathlands, these habitats have dynamic environments; population densities may fluctuate considerably. This toad is an opportunistic breeder, making use of temporary, shallow waters. New habitats can be created rather easily and are quickly taken into use. Although populations in the coastal sand dunes remain stable, threats are posed by new ways of building, resulting in habitats on building sites not being available long enough for successful reproduction. Elsewhere, the bare soil in their terrestrial habitat disappears by vegetational succession. In general terms, the fragmentation of the landscape results in populations becoming isolated from each other.

III. Conservation measures and monitoring programmes

A. Conservation measures

The government is responsible for nature conservation and developing policies for that purpose. Governmental and non-governmental organizations carry out measures in the field, work that is often done by volunteers. The state subsidizes projects both for species and habitats. For some species, there have been national Action Plans. The National Nature Policy Plan, *Natuurbeleidsplan* (Ministerie, 1990), provides the basis for the *Ecologische Hoofdstructuur* (EHS) [National Ecological Network], which consists of core areas, nature development areas, and corridors. This network is still under construction and planned to be completed by 2018. There are 11 species that have been targeted: *Salamandra salamandra, Ichthyosaura alpestris, Triturus cristatus, Lissotriton helveticus, Alytes obstetricans, Bombina variegata, Pelobates fuscus, Bufo calamita, Hyla arborea, Rana arvalia,* and *R. lessonae*. Under the auspices of the Habitats Directive of the European Union, The Netherlands is contributing to the European network of Natura2000 areas by designating protected areas. Amphibians, however, play a minor role in this process.

The main organizations concerned with nature conservation that also own land are *Staatsbosbeheer* (the state forestry service), *Natuurmonumenten, Provinciale Landschappen,* one for each of the twelve provinces, and the *Waterschappen* (the water boards). In addition, the *Stichtingen voor Landschapsbeheer* organize landscape management for each province, focusing mainly on small landscape elements; these organizations do not own land. All organizations carry out management work that includes paying special attention to amphibians and their habitats.

Volunteers play an important role in the conservation of amphibians in The Netherlands. Numbering several thousand and organized into working groups, they conduct surveys, carry out management measures, monitor the situation in the field, and send in distributional data to the national database, which at present comprises some 440,000 records. Together with the RAVON foundation (Reptile, Amphibian & Fish Conservation Netherlands) that coordinates their activities, they are also engaged in promoting interest in amphibians among the general public.

Action Plans can be a useful tool for amphibian conservation. Such a plan for an amphibian species consists of finding out the number of its populations and their sizes, as well as the characteristics of the aquatic and terrestrial habitats and any connections between them. Working with the idea of 'basic habitats', sites were chosen where they could be developed, and the costs estimated. Depending on the resources available, plans were carried out, mostly phasing the work over several years. Once work had been completed, the habitats were monitored to follow their development and to see whether amphibians accepted them. Action Plans have been carried out countrywide for *Alytes obstetricans, Bombina variegata, Hyla arborea,* and *Pelobates fuscus*; Action Plans carried out by individual provinces dealt with additional species of amphibians.

The loss of ponds in the agricultural landscape was a major cause of the decline of amphibians. Thus, plans for ponds were developed that provided for the creation and restoration of waters for use as habitat for reproduction (Boothby 1999). The first pond project started in the province of Limburg in the early 1980s; other provinces followed, and as a result new ponds were dug countrywide. Unfortunately it was often not possible to create ponds at places with the best potential. Furthermore, there was no guarantee of continued maintenance. At first, only the common species benefited from the new and restored ponds; the rare and threatened amphibians for which they were meant did not. Later, measures became more tailored to the requirements of these rare species with the result that their populations were able to build up again locally, and extension of their distributional boundaries was made possible.

Conservation measures in the terrestrial habitat of amphibians are directed largely towards maintaining scrub patches, hedgerows, and other small elements of the landscape, such as heaps of stones and log piles, that can serve as refugia (Stumpel 1997). Limiting the growth of woody vegetation is also necessary for the upkeep of the vegetation mosaic. Among the special provisions for the movement of fauna are ecoducts, tunnels, and specifically-designed kerbstones, the last two more particularly for amphibians.

During the spring migration of amphibians, many volunteers help prevent large numbers ending up as traffic victims by closing off roads at dusk. Where barriers are not possible, warning signs are sometimes put up ordering vehicles to slow down. Mortality is reduced by bordering roadsides with fences containing pitfall traps along its length, so that amphibians can be captured and carried over the road in buckets.

After habitat has been restored, or new habitat created, a species may be reintroduced into an area to help build up a population. This has been done successfully for *Hyla arborea* and *Bombina variegata* by releasing both larvae close to metamorphosis and juveniles. A new approach to conservation is the building up of a breeding stock from threatened populations for releases in new and restored habitats. These habitats can either not be reached by the amphibians in a natural way due to barriers in the landscape preventing their migration, or because the tenacity of a species to a site is so strong that they do not colonize new habitats readily. This stock is obtained by collecting eggs in the field and rearing the young in the laboratory; when the larvae are near metamorphosis, they are set free in the target habitat. After successful projects with *Hyla arborea*, this is now being practiced with *Pelobates fuscus*, taking animals from sites where the species was threatened with extirpation.

Amphibians have also been released outside their natural range by people wishing to extend the species' geographic distribution; the newt *Ichthyosaura alpestris* and the toads *Alytes obstetricans* and *Bufo calamita* are cases in point. The introduced species have persisted for decades and may have affected local populations of other amphibian species. Moreover, by confusing the original distributional pattern, and in some cases because the animals are not of indigenous stock, such introductions hamper research into the biogeography and genetics of native amphibians.

Exotic species have also become established as a consequence of releases. They often pose a threat to indigenous amphibians by competition, predation, and hybridization or because they sometimes carry diseases. This is the case with *Triturus carnifex*, which inhabits the same habitats as the indigenous *T. cristatus* and hybridizes with it. *Rana (Lithobates) catesbeiana* is competing for habitat with indigenous green frogs (*Rana* synkl. *esculenta*) and furthermore is a vector for chytrid infections (Bai *et al.* 2010).

Informing and advising nature wardens, farmers, and the general public forms an important part of nature conservation in The Netherlands. In this, the NGO RAVON plays a central role by creating platforms with all interested parties around the table, giving talks, developing websites, publishing magazines and folders, and keeping in contact with the volunteers in working groups all over the country. Not only has this organization created awareness of the need for protecting amphibians, it also manages the data bank with all records of amphibians, has an extensive documentation centre and, together with students from universities and technical schools, carries out scientific research.

B. Monitoring programmes

In 1997, *Meetnet Amfibieën*, the national amphibian-monitoring network was set up (Goverse *et al.* 2006). Hundreds of well-instructed volunteers (Groenveld *et al.*, 2011) have since surveyed 338 areas that included over 2,500 breeding sites. Monitoring mostly takes place four times a year and this network is still growing. From the amphibian counts, trends in population development are calculated using a computer programme (TRIM) designed for monitoring data with missing values (Pannekoek & van Strien 1998). The status of all species over this period has now been assessed, and shows a moderate increase or stable situation for twelve of the sixteen indigenous species. Although Witmer *et al.* (2002) found a continuing decline in amphibians during the second half of the 20th century; some species seem to be making a recovery during the past decade.

Short-term monitoring also occurs, usually for a period of five years. This is sometimes on the authority of the government that also funds it, or a volunteer working group might be interested in following the development of a local population.

IV. Perspective

Half of all amphibian species in The Netherlands are at present threatened in some way or another. It remains uncertain whether amphibians in general are recovering from a very rough time in the past. Numerous factors plead against this. Most habitats are suffering from drying out as a result of large-scale drainage and from eutrophication and acidification due to airborne nitrate pollution both from agriculture and traffic. Although nature reserves have been designated and new habitats created, many of them are too small and isolated to be able to sustainably support a population. Possible effects of climatic change have not yet been taken into account and remain open to interpretation. Evidence for causal relationships between amphibian decline and other environmental factors is largely lacking, thereby weakening public debate. Modern society is focused on short-term economic profit, which is always at the cost of nature. The fate of the amphibians is in the hands of politicians; only they have the power to turn the tide. It is hoped that the present volume convinces them that effective conservation of amphibians is urgently needed.

V. Acknowledgements

Thanks are due to Jeroen van Delft, Edo Goverse, Claire Hengeveld, Annemarieke Spitzen-van der Sluijs, and Suzette Stumpel-Rienks for comments on the manuscript.

VI. References

Backes, C. and Verschuuren, J. (eds), 2001. *Natuurbeschermingsrecht 2001/2002. Richtlijnen, verordeningen, verdragen, jurisprudentie, Nederlandse wet- en regelgeving en toelichtingen, met name rond de Vogelrichtlijn (79/409/EEG), de Habitatrichtlijn (92/43/EEG) en de CITES-verordening (338/97/EG)*. Sdu publishers, The Hague.

Bai, C., Garner, T. W. J. and Li, Y., 2010. First evidence of *Batrachochytrium dendrobatidis* in China: discovery of chytridiomycosis in introduced American bullfrogs and native amphibians in the Yunnan province, China. *EcoHealth* **7**: 127–134.

Bal, D., Beije, H. M., Fellinger, M., Haveman, R., van Opstal, A. J. F. M. and van Zadelhoff, F. J., 2001. *Handboek Natuurdoeltypen*. Expertisecentrum LNV, Wageningen. Second, fully revised edition.

Boothby, J. (ed), 1999. Ponds and pond landscapes of Europe. *Proceedings of the International Conference of the Pond Life Project, Vaeshartelt Conference Centre, Maastricht, The Netherlands, 30th August–2nd September 1998*. John Moores University, Liverpool.

Bosch, J., Martínez-Solano, I. and García-París, M., 2001. Evidence of a chytrid fungus infection involved in the decline of the common midwife toad (*Alytes obstetricans*) in protected areas of central Spain. *Biological Conservation* **97**: 331–337.

Creemers, R. C. M. and van Delft, J. J. C. W. (eds), 2009. *De Amfibieën en Reptielen van Nederland*. Nederlandse Fauna, Series No. **9**. "Nationaal Natuurhistorisch Museum" Naturalis, European Invertebrate Survey - Netherlands, Leiden.

Goverse, E., Smit, G. F. J., Zuiderwijk, A. and van der Meij, T., 2006. The national amphibian monitoring program in the Netherlands and NATURA 2000. Pp. 39–42 in *Herpetologia Bonnensis. II. Proceedings of the 13th Congress of the Societas Europaea Herpetologica*, ed by M. Vences, J. Köhler, T. Ziegler and W. Böhme. Museum Alexander Koenig, Bonn.

Groenveld, A., Smit, G. and Goverse, E., 2011. *Handleiding voor het monitoren van amfibieën in Nederland*. RAVON Working Group Monitoring, Amsterdam. Third edition.

Ministerie van Landbouw, Natuurbeheer en Visserij, 1990. *Nationaal Natuurbeleidsplan*. Sdu publishers, The Hague.

Pannekoek, J. and van Strien, A. J., 1998. TRIM 2.0 for Windows (Trends and Indices for Monitoring data). Statistics Netherlands (CBS), Voorburg. Research paper No. **9807**.

Spitzen-van der Sluijs, A. M., Zollinger, R., Bosman, W., van Rooij, P., Clare, F., Martel, A. and Pasmans, F., 2010. *Batrachochytrium dendrobatidis* in amphibians in the Netherlands and Flanders (Belgium). RAVON, Nijmegen.

Stumpel, A., 1997. Amphibians and reptiles in agricultural and urban landscapes in the Netherlands; design of and provisions for sub-habitats and corridors. Pp. 53–64 in *Opportunities for Amphibians and Reptiles in the Designed Landscape*, ed by R. Bray and T. Gent. English Nature, Peterborough. English Nature Science Series No. **30**.

Stumpel, A. H. P., 2004. *Reptiles and Amphibians as Targets for Nature Management*. Thesis Wageningen University, Wageningen.

van Delft, J. J. C. W., Creemers, R. C. M. and Spitzen-van der Sluijs, A. M., 2007. *Basisrapport Rode Lijst Amfibieën en Reptielen volgens Nederlandse en IUCN criteria*. RAVON, Nijmegen.

Witmer, M., Dirkx, J., Leneman, H., Notenboom, J., van Veen, M. and Sollart, K. (eds), 2002. *Natuurbalans 2002*. Kluwer, Alphen aan den Rijn.

35 Amphibian declines and conservation in Belgium

Gerald Louette and Dirk Bauwens

Abbreviations and acronyms used in the text and references:

EEC	European Economic Community
EU	European Union
NGO	Non-governmental organization

I. Introduction

This chapter deals with Belgium, a federal state in Western Europe. The country covers an area of some 30,500 km² and is densely populated (*ca.* 360 citizens per km²) and highly industrialized. Belgium is geographically, linguistically, and politically subdivided into two main regions. The region of Flanders (*ca.* 13,500 km²) in the north is home to the Dutch-speaking community. It is a lowland area (average elevation *ca.* 40 m) that experiences an Atlantic climate. Wallonia, the southern part of Belgium (*ca.* 17,000 km²), is inhabited by the French-speaking community and comparatively is an upland (mean elevation *ca.* 310 m) and hilly region, experiencing a more Continental climate. The Brussels-Capital Region, a third autonomous Region, is an officially bilingual enclave situated within the region of Flanders.

II. Legal protection and current status of amphibians

Conservation, including management of nature, forest areas, and associated legislations, are part of the political competencies that have been transferred from the Federal to the Regional Governments. In Flanders, all native amphibians gained legal protection by the Royal Decree of 22 September 1980. This legislation, as well as the Bern Convention and the European Union's Habitats

Directive (Council Directive 92/43/EEC on the Conservation of Natural Habitats and of Wild Fauna and Flora), was implemented in the Ministerial Order of the Flemish Government on 'the Protection and Management of Species' of 1 September 2009. In Wallonia, the Law on Nature Conservation of 13 July 1973, as well as the Bern Convention and the EU's Habitat Directive, were implemented in the Decree of 26 December 2001. This offers legal protection to all native amphibian species and their habitats in Wallonia, except for *Bufo bufo* and *Rana temporaria*. In the Brussels-Capital Region, all amphibians are legally protected (Law of 29 August 1991).

Since the 1970s, considerable efforts have been made by a large group of volunteers to collect distributional data on the native amphibians. This resulted in two atlases that provide detailed and thorough summaries of species' distributions on a fine scale (4 x 4 km grid squares) for Flanders (Bauwens and Claus 1996) and Wallonia (Jacob *et al*. 2007). These publications also include a regional Red Data List for the herpetofauna, based on a combination of extent of the distribution and the degree of recent changes in distributional area or population size. The List was updated for Flanders by Jooris *et al*. (2012). Table 35.1 summarizes both these Red Data Lists.

Despite their legal protection for several decades, there has been a steady decline of many amphibian species in both regions. Below, those species that are of major concern are discussed and the conservation measures that attempt to ameliorate their conservation status are described.

Table 35.1 Status of native amphibians in Belgium, based on Red Data Lists for Flanders (northern Belgium) and Wallonia (southern Belgium). In both regions, status was assessed by the combination of two criteria: the extent of the distribution and the rate of decline. Precise criteria are given by Jacob *et al*. (2007) and Jooris *et al*. (2012).

RE = regionally extinct   **CR** = critically endangered   **EN** = Endangered   **VU** = Vulnerable
NT = Near Threatened   **LC** = Least Concern   **Boldfaced status indicates some measure of threat.**

Species	FLANDERS			WALLONIA		
	Distribution	Decline	Status	Distribution	Decline	Status
CAUDATA						
Lissotriton helveticus	Common	None or weak	LC	Common	None or weak	LC
Lissotriton vulgaris	Very common	None or weak	LC	Common	None or weak	LC
Ichthyosaura alpestris	Very common	None or weak	LC	Common	None or weak	LC
Salamandra salamandra	Very restricted	Strong	**VU**	Common	None or weak	LC
Triturus cristatus	Restricted	Weak	**VU**	Very restricted	Strong	**EN**
ANURA						
Alytes obstetricans	Very restricted	Strong	**EN**	Common	None or weak	LC
Bombina variegata	Extinct	Very strong	**RE**	Very restricted	Very strong	**CR**
Bufo bufo	Very common	None or weak	LC	Common	None or weak	LC
Bufo calamita	Restricted	Weak	**VU**	Very restricted	Strong	**EN**
Hyla arborea	Very restricted	Very strong	**CR**	Extinct	Very strong	**RE**
Pelobates fuscus	Very restricted	Very strong	**CR**	Extinct	Very strong	**RE**
Pelophylax kl. esculentus	Very common	None or weak	LC	Common	None or weak	LC
Rana arvalis	Restricted	Weak	**VU**	---	---	Absent
Rana temporaria	Very common	None or weak	LC	Common	None or weak	LC

III. Species of special conservation concern

A. The yellow-bellied toad (*Bombina variegata*)

In Flanders, this species was historically restricted to the extreme eastern part (Voeren), where it was last observed in the early 1980s. In Wallonia it was once common, especially in the valleys of the river Meuse and some of its effluents (Lesse, Ourthe, Vesdre, Amblève, Warche). According to Boulenger (1922), a remarkable decline in the size and number of its populations was observed as early as about 1880. This decline continued throughout the 20th century. Parent (1997) gave a detailed account of population regression and disappearance from various regions. Actually, only three very small populations survive (Jacob *et al.* 2007). Actions for protecting and extending these populations were started. It is noteworthy that declines of similar magnitude have been observed in neighbouring areas of the Netherlands, Germany, Luxemburg, and France. A multitude of causes for this decline have been named, most of them without much quantitative support.

B. The treefrog (*Hyla arborea*)

The Treefrog formerly had a very large distributional area in both Flanders and Wallonia. By 1975, however, it had disappeared from most locations. The decline was most dramatic in Wallonia, where it was still present in 24 grid squares during the period 1975–1984, but is now considered as extirpated (Jacob *et al.* 2007). As the species is still present in Luxemburg, near the frontier of Wallonia, collaboration with this country was recently established to extend the existing populations.

During the period 1975–1984, treefrogs were present in four distinct and mutually separated areas in Flanders (Bauwens and Claus 1996). In the extreme northern part of the province of West-Flanders, adjacent to the border with The Netherlands, several small populations persisted. In the northern part of the province of Antwerp, where formerly numerous populations existed, only two of them survived. A rather large group of probably interconnected populations was present in the central part of the province of Limburg; at least two populations were composed of > 30 individuals. Finally, three small populations were present in the valley of the river Meuse, in the extreme eastern part of the province of Limburg. During the years 1985–1994, all populations in the province of Antwerp and many of those from Limburg became extirpated. For some populations the decline was well documented and occurred astoundingly fast. For instance, at one locality, a former fish-rearing pond, the number of calling males decreased from about 150 during 1987 to just two in 1990.

In response to this alarming situation, a species protection plan was drawn (Vervoort and Goddeeris 1996) and suggested actions were undertaken, generally without much success. During 2000–2005 another two of the largest populations in Limburg disappeared. At about the same time, small numbers of treefrogs were rediscovered at two nature reserves where they had not been seen for at least 20 years (Lewylle *et al.* 2008). Subsequently, management of these areas was altered to favour the treefrogs. In one reserve, temporal drainage cycles were restored in abandoned fish-rearing ponds. In the other area, 25 pools with strongly fluctuating water levels (dry in autumn) were created. These management practices attempt to eradicate (exotic) fish species that are thought to have direct or indirect negative effects on the eggs and larvae of the treefrogs. Success of the altered management policy was overwhelming. In the first reserve, the central population was estimated to harbour 250–300 calling males in 2010, while two adjacent smaller ponds contained 30 and 25 males. In the second area, all newly excavated pools were colonized by a total of about 350 calling males in 2010 (Engelen and Lewylle 2010).

C. The common spadefoot (*Pelobates fuscus*)

The spadefoot toad is without doubt the most enigmatic amphibian species in Belgium. Due to its secretive habits, including long periods of underground activity and its practice of calling from beneath the surface of the water, it is hard to document its presence and to study its ecology. The preferred habitat seems to consist of rather nutrient-rich pools and dry, sandy and nutrient-poor terrestrial biotopes. This peculiar combination of these rather diverse habitats is very uncommon in Belgium.

In Wallonia, the species was known from a single locality where observations on one adult and several larvae were made in the years 1981–1987. No observations have been recorded subsequently and the population is now considered to have been extirpated (Jacob *et al.* 2007). In Flanders, the species was found in nine grid squares during the period 1975–1995 (Bauwens and Claus 1996). All these localities, except one, were clustered roughly in three groups, all situated in the province of Limburg. During a recent survey (2009) the presence of the spadefoot toad could be confirmed at only four localities (Lewylle and Roossen 2009).

A species-protection plan was drawn up by both the Agency for Nature and Forests and the community of Peer (Roosen 2008). These list an array of very specific and detailed actions including management of potential terrestrial habitats, temporary cleaning of existing pools and construction of new ones, subsidence schedules for farmers, strategies for raising public awareness, and monitoring of the toad populations. Some of these proposed actions have been implemented; results of the monitoring initiatives are not yet available.

D. The common midwife toad (*Alytes obstetricans*)

The midwife toad has a clear preference for habitats with a rather warm microclimate. Hence, it is often found in hilly areas, on exposed surfaces on southward-oriented slopes that have a rocky or stony substrate. The species is often associated with man-made habitats, such as quarries, graveyards, ruins, and other buildings.

The northwestern limit of the species' distribution runs through Belgium and roughly coincides with the border between Wallonia and Flanders. In Wallonia, the species has an ample distribution and populations are generally large. The global population status is considered as being 'stable'. However, populations close to the Walloon-Flemish border seem to be in some form of decline (Jacob *et al.* 2007). In Flanders, the midwife toad is restricted to several small populations at the periphery of its distributional range in Wallonia. A diffuse group of four to five isolated and small populations is found in, or near, the large forest complexes south and southeast of the Brussels-Capital Region. A second group of populations is situated in Voeren, the extreme eastern part of the province of Limburg, with neighbouring populations in the provinces of Liège (Wallonia) and Zuid-Limburg (the Netherlands). Between 1980 and 2001, larvae of the midwife toad were found there in 23 pools, which together formed a seemingly durable (meta)population. Finally, an isolated population is present in the community of Borgloon.

In several of these populations, specific protection actions have been implemented (Vervoort 1994), often by a consortium of organizations, including the Agency for Nature and Forests, Provincial Authorities, the Flemish Land Agency, several 'Regionale Landschappen' and the NGO 'Natuurpunt'. At these localities, a notable increase in population size has been observed, at least temporarily. A summary of actions taken, including photographic impressions, was given by Engelen and Jooris (2009). By contrast, a recent survey at Voeren, which was formerly considered a stronghold for this species, yielded alarming results. Larvae were found in only two pools and only six males were heard calling (Engelen and Jooris 2009). Specific actions are now being taken to recover these populations.

E. The fire salamander (*Salamandra salamandra*)

In Belgium, the Fire Salamander exhibits a clear preference for deciduous forests, most often in areas with a marked relief. They deposit their larvae in wells and small brooks with clear, oxygen-rich water, although occasionally also in stagnant waters of small pools or even in car tracks on forest roads. The northwestern limit of the species' distribution runs through Belgium and roughly coincides with the border between Wallonia and Flanders.

In Wallonia, the species has a widespread distribution and populations are generally large, especially south of the rivers Meuse and Sambre. The global population status is considered as being 'stable'. However, populations close to the Walloon-Flemish border occupy fragments of forest, are mutually isolated, and some have become extirpated due to alterations of aquatic habitats (Jacob et al. 2007). In Flanders, the fire salamander is restricted to localities at the periphery of its distributional range in Wallonia. All localities, except one, are situated in the provinces of East-Flanders and Flemish-Brabant and correspond to small (< 5 ha) or medium-sized (up to 200 ha) forest fragments. These are the remnants of the large deciduous forests that covered these areas until the Middle Ages, when extensive exploitation of forests began. Due to progressive deforestation and destruction of aquatic habitats, the number of known localities declined strongly in the period 1970–1994. A recent survey evidenced the persistence of fire salamanders in a considerable number of the forest fragments, despite their mutual isolation and the small size of favourable habitat (Jacobs 2009). A species-protection plan (Jacobs 2008) lists several small-scaled and specific management actions to improve the status of local populations. Many of these actions pertain to the management or creation of aquatic habitats. Where implemented, fire salamanders seem to have responded quickly to these actions. Although many populations remain highly vulnerable due to their small size and to the fragility of the habitats, prospects for their long-term survival nowadays seem better than they were about 15 years ago.

F. The natterjack toad (*Bufo [Epidalea] calamita*)

The Natterjack Toad used to have a widespread distribution in Flanders, where it mainly occurred on dry and sandy soils. Nowadays, it is restricted to the western part of the coastline, anthropogenically elevated terrains near the harbour of Antwerp, and the main heathland areas in the Campine ecoregion (Bauwens and Claus 1996). The decline of the species is primarily attributed to loss of habitat. Over the years the extensive landscape of Flanders changed in favour of grey infrastructure in combination with an intensified agricultural activity. In Wallonia, the species was once a common member of the amphibian fauna but now only occurs in five distinct clusters, those being central Hainaut, the environs of Liège, east Fagne, west Famenne, and a military domain in Lorraine. Hence, the species is considered as endangered and an action plan is elaborated. Forestation of formerly open areas, such as meadows and heathlands, has led to the decline of the species, especially in the Ardennes ecoregion (Jacob et al. 2007).

Current industrial and recreational activities seem, on first sight, to be beneficial for the species. Preparatory large-scale supplementations of sand near the Antwerp port led to an exponential increase of the species in that specific area. However, as the port itself is now expanding, it is clear that the species needs to make room. Efforts are made to maintain this pioneer species (listed on Annex IV of the Habitats Directive) by applying the principle of ecological infrastructure. Core populations are connected with satellite populations (so-called habitat backbone concept) in the prevalent mosaic of docks, container terminals, and large-scale hardened structures (Snep and Ottburg 2008). Immediate action is sometimes needed in such dynamic developments of ports and grey infrastructure. Instant disappearance of pools used for reproduction has literally demanded the translocation of the offspring of natterjack toads to safer environments.

G. The great crested newt (*Triturus cristatus*)

The Great Crested Newt is the largest newt inhabiting Belgium. It typically favours permanent ponds with moderate to abundant submerged macrophytes and few, or no, fish. Its terrestrial habitat involves small landscape elements, such as hedgerows, tree rows, and scrub. Before the intensification of agriculture (leading to the loss of habitats) and the infilling of grey infrastructure (leading to isolation), this species had a widespread distribution in Belgium. Nowadays, it is still present in all provinces of Flanders, but scattered over very fragmented populations. The main clusters are near the coastline, southwest of West-Flanders, Antwerp Campines, and punctuated along some rivers (e.g. Schelde, Dijle, Demer, Gete) (Bauwens and Claus 1996). The same holds true for Wallonia where the species is declining but still present in all major ecoregions, except for the forested Ardennes (Jacob *et al.* 2007).

Because of its flagship status as a typical species of agricultural land, as well as its listing on Annex II of the Habitats Directive, this species gets a lot of attention in species-protection programs. Restoration of farmland ponds is a popular measure by local authorities but does not always lead to the desired results. The often-isolated character of these interventions (focusing on a very local scale) needs in many cases a more regional approach. Dissolving dispersal barriers, overall enhancement of environmental quality (of both aquatic and terrestrial habitat), and probably assisted re-colonization are sometimes the only solution for a successful recovery. A good example of such an integrative approach was performed in Wetteren, where a co-operation of the municipality, the Agency for Nature and Forests, the NGO 'Natuurpunt', and local landowners developed a large sustainable habitat patch in favour of newt species. Several ponds were restored (deepened, fish removed, incoming agricultural canals deviated) and new ones created, and the surrounding landscape matrix was adjusted (construction of hedgerows and scrub zones) (Verbelen and Jooris 2009). Yearly monitoring of the project currently indicates good results. In both Flanders and Wallonia large-scale projects supported by the European Union (e.g. Life projects) were launched and aim at the restoration of the species and its habitat. The success of these actions has to be evaluated in the future.

IV. Conservation and monitoring

The legislative framework of species conservation in Flanders (Order on the Protection and Management of Species, 2009) and Wallonia (Law on Nature Conservation, 2001) allows, in theory, a swift design and implementation of species-protection programmes. However, during the past decade such programmes were set-up for only a few species in Flanders (i.e. *Alytes obstetricians, Hyla arborea, Pelobates fuscus, Salamandra salamandra*) and for the three most endangered species in Wallonia (*Bombina variegata, Bufo [Epidalea] calamita, Triturus cristatus*). Although some of these plans were launched too recently to evaluate their success, others rarely have led to successful results, mainly because of brief and often inadequate restoration measures. Large-scale approaches, such as integral habitat restoration, specific management of predators and competitors, and serious enhancement of dispersal routes, are badly needed. Lack of support from landowners and land-managers, in combination with limited financial inputs inadequate for realization, are primarily responsible for these failures. As for a number of species (e.g. *Pelobates fuscus*) for which urgent conservation is necessary, additional actions need to be implemented. Thoughts are ongoing to scientifically guide the reintroduction of this species in its former, but now restored, habitat in the Campine ecoregion (Flanders).

The presence of alien invasive amphibians is an issue that also afflicts Belgium's herpetofauna. Expanding populations of the American bullfrog (*Lithobates catesbeianus*) (Jooris 2005) are thought to transfer the chytrid fungus (*Batrachochytrium dendrobatidis*) and ranaviruses and thereby form a serious threat to indigenous species. Several species of the 'green frogs' group, initially the marsh frog (*Pelophylax ridibundus*) and more recently also the Levantine frog (*P. bedriagae*) and the Anatolian marsh Frog (*P.* cf. *bedriagae*) have been introduced into several localities and are now expanding (Holsbeek 2010; Jooris and Holsbeek 2010). These species may hybridize with related *Pelophylax lessonae* (listed on Annex IV of the Habitats Directive) and disturb the genetic integrity of *P. lessonae* populations. To counteract the negative effect of alien amphibians, Flanders has launched initiatives for their control. For instance, efforts are made to remove American bullfrogs by direct trapping with fyke nets, with encouraging results (Louette *et al.* 2013). Indirect actions are performed by altering their reproduction sites to more ephemeral and fishless habitats, these being the optimal habitat for reproduction for the majority of native amphibians (Louette 2012).

In conjunction with amphibian conservation, monitoring is essential to get a picture of the different species' status. Building further on the already ongoing NGO-driven observation network of volunteers, the government of Flanders has the ambition to develop and support a rigid monitoring program. At least for the species listed in the Habitats Directive, detailed data on status and trends for both range and population will be acquired. In Wallonia a monitoring program of the most endangered species (e.g., *Bombina variegata*, *Bufo* [*Epidalea*] *calamita*) is also being developed. Results from this monitoring will allow better protection of the different species by a regular update of their Red List status, accompanied by development of species-conservation programmes. In addition to the governmental plan of this regionally based monitoring, site-specific monitoring programmes are nowadays established, most often after restoration or creation of local habitat has been initiated. For instance, the restoration of bodies of water for reproduction by *Hyla arborea* during the past five years (Vijvergebied Midden-Limburg and Brand-Maaseik) displayed promising effects on population size. Prolonged monitoring is, however, not structural and therefore depends on the enthusiasm of local volunteers.

V. Acknowledgments

We thank Christiane Percsy and Robert Jooris for their thoughtful comments on a previous version of the manuscript.

VI. References

Bauwens, D. and Claus, K., 1996. Verspreiding van amfibieën en reptielen in Vlaanderen. De Wielewaal, Turnhout.

Boulenger, G. A., 1922. Quelques indications sur la distribution en Belgique des batraciens et reptiles. *Les Naturalistes Belges* **3**: 52–53, 71–77.

Engelen, P. and Jooris, R., 2009. Actuele status van habitatrichtlijnsoorten in Vlaanderen: de vroedmeesterpad. *Hyla-Flits* **1**: 3–7.

Engelen, P. and Lewylle, I., 2010. Boomkikker in Vlaanderen: terug van weggeweest. *Hyla-Flits* **2**: 1–2.

Holsbeek, G., 2010. *Sneaky invasions in European Waterfrogs*. PhD thesis, K.U. Leuven, Leuven.

Jacob, J.-P., Percsy, C., de Wavrin, H., Graitson, E., Kinet, T., Denoël, M., Paquay, M., Percsy, N. and Remacle, A., 2007. Amphibiens et reptiles de Wallonie. Publication d'Aves – Raînne, Gembloux.

Jacobs, I., 2008. Toestand van de vuursalamander in Oost-Vlaanderen. Ecologie, verspreiding en aanzet tot soortbescherming. Rapport Natuurpunt Studie 2008/7, Mechelen.

Jacobs, I., 2009. Vuursalamanders in Oost-Vlaanderen. *Natuur.focus* **8**: 128–134.

Jooris, R., 2005. De Stierkikker in Vlaanderen. Nieuwe inzichten in verspreiding, foerageergedrag en ontwikkeling. *Natuur.focus* **4**: 121–127.

Jooris, R. and Holsbeek, G., 2010. 'Groene kikkers in Vlaanderen en het Brussels Hoofdstedelijk Gewest'. Rapport Natuur.studie, Mechelen.

Jooris, R., Engelen, P., Speybroeck, J., Lewylle, I., Louette, G., Bauwens, D. and Maes, D., 2012. De IUCN Rode Lijst van de amfibieën en reptielen in Vlaanderen. INBO.R.2012.22, Brussel.

Lewylle, I. and Roosen, R., 2009. Knoflookpad in Limburg: het tij gekeerd? *Hyla-Flits* **2**: 1–2.

Lewylle, I., Goddeeris, B., Engelen, P., Roosen, R., De Becker, P. and Herremans, M., 2008. De boomkikker op een keerpunt? Soortgericht beheer boekt eerste resultaten. *Natuur.focus* **7**: 84–93.

Louette, G., 2012. Use of a native predator for the control of an invasive amphibian. *Wildlife Research* **39**: 271–278.

Louette, G., Devisscher, S. and Adriaens, T., 2013. Control of invasive American bullfrog *Lithobates catesbeianus* in small shallow water bodies. *European Journal of Wildlife Research*.

Parent, G.H., 1997. Contribution à la connaissance du peuplement herpétologique de la Belgique – Note 10. Chronique de la regression des batraciens et reptiles au cours du XXieme siècle. *Les Naturalistes Belges* **78**: 257–304.

Roosen, R., 2008. Soortbeschermingsplan knoflookpad (*Pelobates fuscus*). Limburgs Landschap in opdracht van ANB, Heusden-Zolder.

Snep, R. P. H. and Ottburg, F. G. W. A., 2008. The 'habitat backbone' as strategy to conserve pioneer species in dynamic port habitats: lessons from the natterjack toad (*Bufo calamita*) in the Port of Antwerp (Belgium). *Landscape Ecology* **23**:1277–1289.

Verbelen, D. and Jooris, R., 2009. Lange strijd voor populatie kamsalamander in Wetteren. *Hyla-Flits* **2**: 4–5.

Vervoort, R., 1994. Soortbeschermingsplan voor de vroedmeesterpad (*Alytes obstetricans*) in Vlaams-Brabant. KBIN in opdracht van AMINAL Natuur, Brussel.

Vervoort, R. and Goddeeris, B., 1996. Maatregelenprogramma voor het behoud van de boomkikker (*Hyla arborea*) in Vlaanderen. KBIN in opdracht van AMINAL Natuur, Brussel.

36 Amphibian declines and conservation in France

Jean-Pierre Vacher and Claude Miaud

Abbreviations and acronyms used in the text and references:

ACEMAV	*Association pour la connaissance et l'étude du monde animal et végétal*
ASL	*above sea level*
Bd	*Batrachochytrium dendrobatidis*
EC	*European Community*
MNHN	*Muséum national d'Histoire naturelle*
NGO	*Nongovernmental organization*
RACE	*Risk Assessment of Chytridiomycosis to European Amphibian Biodiversity*
SAC	*Special Area of Conservation*
SHF	*Société herpétologique de France*
UE (EU)	*European Union*
UICN (IUCN)	*International Union for the Conservation of Nature*

I. Introduction

Metropolitan France is approximately 550,000 km², extending from 41°N to 51°N in latitude (Corsica included), and with a broad range of biogeographical units and categories of habitats. Of the nine biogeographical regions that are defined within Europe according to the EC Habitats Directive (92/43/EEC), four are found in France: Alpine, Atlantic, Continental, and Mediterranean. At a more precise scale, there are (1) five main mountain ranges: French Alps, Pyrenees, Massif Central, Jura, and Vosges (included in the Alpine Biogeographical Region); (2) five main drainage basins: Garonne, Loire, Rhône, Seine (included in the Atlantic Biogeographical Region), and Rhine (included in the Continental Biogeographical Region); (3) three coastal areas: the Atlantic Coast, the Channel Coast (included in the Atlantic Biogeographical Region), and the Mediterranean Coast (included in the Mediterranean Biogeographical Region).

Many different kinds of habitats are found in these units: (1) garrigue and maquis shrublands, respectively west and east of the Mediterranean region; (2) causses (limestone plateaus) between the Massif Central and the Mediterranean, (3) Atlantic plains (Garonne in the south, Loire in the centre, Seine in the north); (4) Atlantic coastal sand dunes along the Atlantic Ocean from Biarritz to southern Brittany, (5) Western bocage (network of hedgerows) located in Poitou-Charentes, Pays-de-la-Loire, Brittany, and Normandy, (6) continental plains in Lorraine and Alsace, and (7) the flatlands in the north (Nord-Pas-de-Calais, Champagne-Ardenne). Within these great units, many other habitat subdivisions are found and described. Because of this great heterogeneity in landscapes and biomes, France is a biogeographical junction and therefore unique in the composition of its amphibian fauna.

France has 34 species[1] of amphibians, with 11 caudates and 23 anurans. Because of the heterogeneity of landscape, habitats, and biogeographical units, five main amphibian assemblages can be defined that occupy distinct types of habitats. These are assemblages characteristic of (1) open land, (2) evolved habitats, (3) montane habitats, (4) anthropogenic habitats, and (5) caves. These assemblages can be divided further into occupants of various divisions and subdivisions of the major habitats. A total of 37 such amphibian assemblages have been defined for mainland France and Corsica (ACEMAV coll., Duguet and Melki 2003) (Table 36.1).

A. Threats to amphibians in France

Even if there is no specific study at the national scale, the pattern observed at the global scale (e.g. Stuart *et al.* 2008) is valid for France. Loss and degradation of habitat are the greatest cause of the disappearance of amphibians. Introduced species and diseases also have to be considered as they can cause sudden and dramatic population declines resulting in rapid extirpation, even in remote areas. Data collected at the national scale for amphibian distribution (Castanet and Guyétant 1989; Lescure *in press*) have not been collected to access the relative importance of one threat compared with another for a particular species. Such information has to be collected even at a regional scale to allow a finer analysis of significant threats to amphibians.

1. Destruction and alteration of habitats

From 1965 to 1995, the surface area of French wetlands was reduced by approximately half. By 1999, it was estimated that they occupied 1.5 to 1.6 million ha, which is about 3% of the whole territory (Barnaud and Fustec 2007). Also, almost 70% out of the 2 million km of hedges presumably present in France during the climax of the bocage from 1850 to 1910 have been destroyed, i.e.

---

1   Following the suggestion of Speybroeck *et al.* (2010), this number does not include *Bufo balearicus* as a distinct species.

1.4 million km (Pointereau 2002). In the same way, more than 2.4 million ha of meadows have been ploughed and transformed into crops between 1975 and 1995 (Poiret 2005). The number of ponds in metropolitan France is currently estimated at around 600,000 (Pôle-relais mares, zones humides intérieures et vallées alluviales, unpublished report). This figure represents 10% of the ponds known to have been present in the country in 1900 and 50% of the ponds in 1950. As aquatic breeding places, the proportion and distribution of landscape structural elements such as hedges, forests, and prairies, are very important for the conservation of amphibians (e.g. Joly *et al.* 2001). Consequently, the abundance and distribution area of many species has been drastically reduced during the second half of the 20th century in France. This is particularly the case for species such as the common spadefoot, *Pelobates fuscus* (ACEMAV coll. Duguet and Melki 2003; Lescure 1984) and the yellow-bellied toad, *Bombina variegata* (Lescure *et al.* 2011).

Even if not completely destroyed, habitat (both breeding and terrestrial) suffers alteration from fragmentation and pollution. Limited population exchanges and colonization, and toxicological impact of pollutants on the various phases of the complex amphibian life cycle contribute to weaken the amphibian community. This observation is likely to remain valid well into the 21st century as many projects are underway (urbanization, highways) and intensive agricultural practices continue to develop.

2. Diseases

The knowledge of disease epidemiology (including parasites) is particularly sparse in France. The fungus *Batrachochytrium dendrobatidis* (*Bd*) has been detected in many European countries (Dejean *et al.* 2010a) and mass mortality of amphibians has been observed in Spain (Bosch *et al.* 2001). Mortality of *Salamandra salamandra* and *Alytes obstetricans* attributed to *Bd* occurred in several alpine lakes in the French Pyrenees and several *Alytes obstetricans* populations from pristine areas are suspected to have disappeared (D. Schmeller personal communication).

Studies on the prevalence and identification (lineages) of *Bd* are in progress in France under the scope of the UE Biodiversa Risk Assessment of Chytridiomycosis to European Amphibian Biodiversity (RACE) programme (2009-2012, M. Fischer, coordinator). It is too early to give a comprehensive pattern of the distribution of *Bd* in France. The current data confirm the presence of *Bd* in introduced species such as the American bullfrog *Lithobates catesbeianus*, the Balkan water frog *Pelophylax kurtmuelleri*, and in native species such as *Ichthyosaura alpestris, Lissotrition helveticus, Hyla arborea,* and *Bufo bufo*.

The RACE programme carried out at the national scale involves several research laboratories that work together with field naturalists from NGOs, nature reserves, National and Regional Parks, and forest and wildlife agencies from all regions. This collaboration enables the gathering of a very large number of samples from across the country and also warns of mass mortality. This study and the tools now available (www.alerte-amphibien.fr) will allow the collection of data on amphibian diseases (including not only *Bd*, but others such as *Ranavirus* and parasites) and prompt reaction following mass mortality.

Another important issue is the provision of information to fieldworkers about the use of standard protocols of hygiene to limit the spread of *Bd*, other potential diseases, and invasive species (Dejean *et al.* 2010b; available on the *Société Herpétologique de France* (SHF) website www.lashf.fr).

3. Non-native species

Several non-native amphibian species are naturalized in France, including the American bullfrog, *Lithobates catesbeianus,* (Ficetola *et al.* 2007) and the African clawed frog, *Xenopus laevis* (Fouquet and Measey 2006). A regional programme on the feasibility of eradicating the American bullfrog was developed in 2000 (Detaint and Coïc 2005) but its implementation failed in southwestern

France. However, the eradication protocol was followed in two other regions where American bullfrog populations are now declining. A similar feasibility study is now being conducted on the African clawed frog in Western France.

The status of water frogs of the genus *Pelophylax* is complex but genetic studies confirm the existence of allochthonous genetic material among native French water frogs. This certainly results from the importation of water frogs from Spain, the Balkans, and Turkey for human consumption (Schmeller *et al.* 2007; Holsbeek *et al.* 2008). The Italian crested newt, *Triturus carnifex*, was introduced into France in the vicinity of Lake Geneva, where it seems to have colonized several marshes. The impacts of these non-native species, which are in a spreading phase, on native amphibians (and biodiversity in general) vary from direct predation and competition to genetic pollution and the transmission of diseases (e.g. Peerler *et al.* 2001).

The introduction of fish is a serious threat to many amphibian species. Many cases are reported for France. In the Yves marsh (Charente-Maritime) for example, the presence of catfish in ponds is negatively correlated with the presence of *Pelobates cultripes* tadpoles (Thirion 2006). In the Pinail Nature Reserve (Charente), the introduction of the pumpkinseed, *Lepomis gibosus*, to ponds leads to the disappearance of most amphibian species (Dubech personal communication). In the Bresse region (eastern France), the abundance and composition of the newt assemblage in ponds depend on the presence-absence of introduced fishes (Joly *et al.* 2001). The stocking of fishes (mostly salmonids) in alpine lakes is still frequent and impacts amphibian assemblages as described for many other countries (Knapp and Matthews 2000; Orizaola and Braña 2006).

Other aquatic organisms, such as crayfish, can also severely threaten amphibians and this is of increasing concern (Semlitsch 2003; Cruz *et al.* 2006). Along the Rhine River in Alsace, the crayfish *Orconectes limosus* feeds on larvae of the European treefrog, *Hyla arborea* (Vacher personal observations). It is not known, though, if this kind of predation has an impact on treefrog populations. In Spain, Cruz and Rebelo (2007) showed that another species of crayfish (*Procambarus clarkii*) has a negative impact on several species of amphibians, including the Mediterranean treefrog, *Hyla meridionalis*.

II. Declining species of amphibians and species of concern for conservation

A. Species with a reduced distribution in France

1. Species that reach their distributional limit in France

A. Salamandra atra

The Alpine salamander, *Salamandra atra*, is present in the European Alps, from the Northwestern Alps to the Balkan Dinaric Alps (most frequent between 800m and 2000m asl). In France, its presence and specific status has been confirmed recently (Ribéron *et al.* 2003). The species was discovered in only one place close to the village of Samoëns in the department of Haute-Savoie. Despite several surveys, no other populations have been found. Due to its very restricted range, close monitoring and further surveys are required.

B. Bufo viridis

The green toad, *Bufo viridis*, reaches its eastern distributional limit in northeastern France, in Alsace and in northern Lorraine. In both regions, the populations are geographically isolated. In Lorraine, it occurs only north of the department of Moselle where it inhabits ancient industrial sites. In Alsace, two main populations are present in the western surroundings of Strasbourg (Bas-Rhin) and in a few localities north of Mulhouse (Haut-Rhin). The species is found there in gravel and sand quarries, old mining sites, and in agricultural landscapes. In Alsace at least, the habitat is highly fragmented by roads (including several motorways), urbanization, and intensive agriculture.

The green toads that occur in Corsica have recently been assigned to *Bufo balearicus* Boettger, 1880 (Stöck *et al.* 2008). However, the designation of *Bufo balearicus* as a true species is debated and some authors recommend treating all *Bufo balearicus* populations as *Bufo viridis* until further taxonomic investigation is conducted (Speybroeck *et al.* 2010). The green toad population of Corsica is not considered endangered and does not require urgent conservation measures advocated for the populations of northeastern France.

C. Pelobates fuscus

The common spadefoot, *Pelobates fuscus*, reaches its eastern distributional limit in France. In the 19th century, it was distributed in 25 departments from the central region south of Paris throughout Champagne-Ardenne, southwestern Picardie, Lorraine and Alsace (Lescure 1984; Eggert 2002; ACEMAV coll. Duguet and Melki 2003). It declined throughout the 20th century and is nowadays only known from five departments: Indre, Loiret, Moselle, Bas-Rhin, and Haut-Rhin. In these, populations only occupy small ranges. In Indre and Loiret, for example, it is only known from one locality in each department. In Moselle, it is restricted to an industrial area in the north of the department. In Alsace (Bas-Rhin and Haut-Rhin), it occurs only in eight localities on the floodplain of the Rhine (Vacher and Dutilleux 2010). The main threats to the common spadefoot are alteration and loss of appropriate aquatic habitats for reproduction and the loss of traditional agricultural practices that provided suitable terrestrial habitats, such as the culture of asparagus in Alsace.

D. Rana arvalis

The moor frog, *Rana arvalis,* is known to occur only in two localities in the Nord department. It used to occur in the Rhine Valley and in the Sundgau in Alsace and Territoire-de-Belfort, but no population has been found there since the late 1970s (Vacher *et al.* 2008, 2010a). Three isolated individuals were found in alluvial habitats along the Rhine in 2006 and 2009 but no evidence of reproduction or of an actual population have been found (Vacher 2010a, 2010b). It is likely that this species is on the verge of extirpation in northeastern France (Alsace and Franche-Comté regions). The moor frog reaches its eastern limit in France and is therefore naturally rare there. The canalization of the Rhine River induced important ecological changes in the alluvial forests and wetlands of this floodplain. Many ponds, marshes, and branches of rivers became overgrown and underwent rapid eutrophication. Canalization of the Rhine also induced a lowering of the ground water. Many humid meadows and small marshes are no longer, or only briefly, flooded during spring and thus it is difficult for this rare species to complete its reproductive cycle and it is declining.

2. Narrowly distributed species that occur in France

A. Calotriton asper

The Pyrenean brook salamander, *Calotriton asper*, is endemic to the Pyrenees. It does not seem to be endangered in France. Due to its restricted range, however, and strict ecological demands, it may become threatened in some parts of the Pyrenees by global climatic change, the introduction of fish into montane streams, and chytridiomycosis.

B. Salamandra lanzai

Lanza's alpine salamander, *Salamandra lanzai*, has a very restricted world range at the border between Italy and France. It is found in the Western Alps in the valleys of Po, Pellice, and Germanasca in Italy, and in the Guil Valley in France. Even though the species does not seem to be particularly endangered nowadays, it is rather vulnerable because of its very restricted range; close

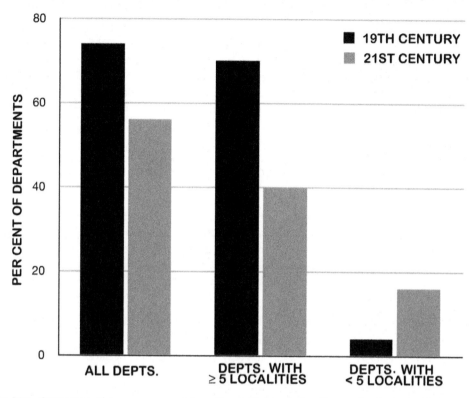

Fig. 36.1 Comparison of the occurrence of the yellow-bellied toad, *Bombina variegata*, in the departments of France between the 19th and 21st Centuries. Data from *Lescure et al* (2011).

monitoring is needed. Possible threats include overcollecting and climatic change (Andreone *et al.* 2004).

C. *Rana pyrenaica*

The Pyrenean frog, *Rana pyrenaica,* is endemic to the Pyrenees and has a rather limited range near the Spain-France border (Gosá Oteiza *et al.* 2010). In France, it is known to occur only in a small area in the department of Pyrénées-Atlantiques. It occurs in a few streams at elevations approximately between 500m and 1000m asl) on the southern slopes of the Iraty Forest (Llamas *et al.* 1998; Berroneau 2010). As far as presently known, the habitats for this species are in good conservation status. Due, however, to its very small geographic range and the possible threats posed by management of forests, introduction of fish, chytridiomycosis, and climatic change, this species is considered as endangered (Comité français de l'UICN 2009). Special conservation efforts such as surveys, monitoring, and further scientific investigation are implemented in the scope of a conservation action plan for this species in order to regularly assess its status.

B. Widely distributed species that occur in France

Even though there are not yet enough data to provide a relevant analysis of the global trends of amphibian populations in France, there is evidence that some rather widespread species are locally declining. For example, Boissinot (2010) showed that common species such as *Salamandra salamandra* and the water frogs of the genus *Pelophylax* are in fact declining in some areas of the department of Deux-Sèvres due to intensification of agriculture leading to homogeneity in types of habitats.

Another example of a quite widespread species with conservation issues is the yellow-bellied toad, *Bombina variegata*. This species is distributed mainly in the eastern half of the country, although it is also found west to Poitou-Charentes, Limousin, and in the north of Midi-Pyrenees (Lescure *in press*). A recent study used distributional data to compare the status of populations before 1900 with that after 2000 (Lescure *et al.* 2011) and showed that the species has disappeared from 24% of the departments from which it was known in the 19th century (Fig. 36.1). Moreover, as of 2011, the number of departments with fewer than five occupied localities is 14, whereas this number was only four before 1900.

III. Conservation measures and monitoring programmes

A. National action plans

In 2009, the French Ministry of Ecology and Sustainable Development launched three national action plans for amphibians: *Bombina variegata, Pelobates fuscus*, and *Bufo viridis*. These plans are expected to be implemented for five years from 2011 to 2015. In the scope of these plans, three main topics are treated: (1) knowledge and research, (2) conservation and management measures, and (3) public awareness. These action plans aim at maintaining or restoring a good conservation status for all populations of these species.

B. The Natura 2000 Network and the Habitats Directive in France

1. Amphibians and the Natura 2000 Network

The Terrestrial Special Area of Conservation (SACs) covers 4.6 million hectares in France, which represent about 8% of the national territory. In each area, the regional department of environment (*Direction régionale de l'Environnement, de l'Aménagement et du Logement*) is in charge of implementing Natura 2000 (the EU protected areas network) together with various partners (e.g., national and regional parks, nature reserves, national forest agency, counties). Prior to implementation in the field, a management plan is written for each SAC. This document includes a precise mapping of site and a complete overview of the ecological issues and describes recommendations for management for each important issue, whether concerning habitats or species. When a SAC has been designated specifically for a species, then recommendations for management focus on this particular species. When a SAC has not been designated for a particular species but still one or more "Annex II" species occur on the site, then conservation actions are to be proposed for these species. There are five Annex II amphibian species in France: (1) *Speleomantes strinatii* (found only in a few localities in the Alpes-Maritimes department); (2) *Triturus cristatus*; (3) *Bombina variegata*; (4) *Discoglossus montalentii* (found only in Corsica); (5) *Discoglossus sardus* (found in Corsica and on the islands of Hyères and Port-Cros).

In the SACs where these species occur, targeted management actions are to be proposed and implemented through Natura 2000 contracts in order to maintain populations in a good conservation status, or to enhance the status if it is considered poor or bad. This network is beneficial in that it enables land managers to dig or restore ponds and to maintain extensive agricultural practices that are suitable for amphibians.

2. Implementation of Article 17 of the Habitats Directive

Within the scope of Article 17 of the EU Habitats Directive, a first evaluation carried out in 2006 by the *Muséum national d'Histoire naturelle* (MNHN) and the SHF showed that 21 out of 25 of the Annex IV species present in France had an unfavourable conservation status. Due to the lack of precise data concerning amphibian populations throughout the whole country, this assessment was made using distributional data gathered for the national atlas of amphibians and reptiles, carried out by the SHF. This evaluation will be updated in 2012.

C. Monitoring of amphibians in France

In 2009, the SHF launched a national monitoring programme named POPAMPHIBIEN (available at www.lashf.fr). This programme aims at defining the global tendency of amphibian populations in the whole country. It is based on the presence-absence and abundance occupancy model statistical framework. In the field, it involves a large network of volunteers that are active in the regional sections of the SHF and other local conservation organizations and also many habitat managers such as the national forest agency (*Office national des Forêts*), nature reserves, and national parks. Agreements are concluded between the SHF and different partners in order to adapt local protocols regarding management issues and their implementation. The SHF also has an agreement with the laboratory of species conservation, restoration, and monitoring of populations of the MNHN that aims at analyzing the data gathered through this protocol.

This programme will produce relevant estimates of trends in amphibian populations at the national scale. Moreover, it also aims to connect observed trends with environmental variables, e.g. assessment whether a population decrease in a managed area is the result of wrong practices or if it is linked with a global phenomenon. The programme will also illuminate the trends in populations of very common species such as the common toad, *Bufo bufo*, the common frog, *Rana temporaria*, and the palmate newt, *Lissotriton helveticus*, for which currently there is no assessment as to whether or not they are declining on a broad scale.

D. Protected areas

1. National parks and nature reserves

The two main types of protected areas in France are national parks and nature reserves. The latter may be national (funded by the State) or regional (funded by regions). Five national parks of metropolitan France are all located in mountainous areas and one is a marine park. The nature reserves are located in many different areas. Each is ruled by a management plan that is implemented for a duration of six years. After that period, the management plan is updated and renewed and implemented for another six years and so on.

2. Other protected areas

Non-governmental nature conservancies are found in each of the 22 regions of France. They are grouped under the national network of nature conservancies (*Conservatoire des Espaces naturels*). Their main goals are to acquire and manage lands for conservation purposes. They can also manage lands that are not their own property but that belong to different local governments or even to municipalities through agreements with these partners. In some regions, e.g., Alsace, conservancies can also manage nature reserves, both national and regional. The conservancy sites have a management plan that lists the measures that should be taken in order to preserve habitats and species. Many of these sites are wetlands and the management plans take into account the requirements needed for amphibian conservation.

In France, the Prefect of a department can designate a protected natural area under an act, these zones are called Prefectural Habitat Protection Orders (*Arrêté Préfectoral de Protection de Biotope*). These areas are usually managed by nature conservancies or non-governmental conservation organizations. Many wetlands with amphibian populations benefit from such a measure.

Finally, each local government in all French departments (*Conseil général*) can buy lands and manage them under the notable natural habitat (*Espace naturel sensible*) policy. This is a conservation tool based on acquiring lands for protection. Once an area is acquired in such a way, it can be managed by the environmental department of local government or the management can be delegated to the natural habitats conservancies.

Table 36.1 Types of habitats occupied by amphibian assemblages in France (from ACEMAVcoll. *et al.* 2003).

Habitats	Divisions	Subdivisions
Open land habitat	4	7
Evolved habitats	4	18
Mountane habitats	4	5
Anthropogenic habitats	0	3
Cave habitats	0	0

IV. The Red List of amphibians of metropolitan France

In 2008, the French committee of the IUCN, together with the MNHN and the SHF, completed the first Red List of amphibians and reptiles of metropolitan France (Table 36.2). This list was first published on the internet in 2008, and as a booklet in hardcopy in 2009 (Comité français de l'UICN 2009). The national Red List for amphibians of metropolitan France provided an up-to-date taxonomic list of amphibians found in France, including non-native species.

Table 36.2 The amphibians of France and their status in the national Red List published by the French committee of the IUCN and MNHN (*Comité français de l'UICN 2009*).

Species marked with an * are introduced and are listed as **NA**= Not Applicable   **DD**=Data Deficient   **LC**= Least Concern.

The categories indicating some degree of threat are colour coded: **NT**=Near Threatened (tan)   **VU**=Vulnerable (yellow)   **EN**=endangered (rust brown)   **CR**=Critically Endangered (red)

For explanation of subcategories, see IUCN and MNHN reports cited above.

SPECIES	RED-LIST CATEGORY
Speleomantes strinatii	NT
Calotriton asper	NT
Euproctus montanus	LC
Ichthyosaura alpestris	LC
Lissotriton helveticus	LC
Lissotriton vulgaris	LC
Salamandra atra	VU D2
Salamandra corsica	LC
Salamandra lanzai	CR B1ab(i,ii,iii,iv)
Salamandra salamandra	LC
Triturus carnifex*	NA
Triturus cristatus	LC
Triturus marmoratus	LC
Alytes obstetricans	LC
Discoglossus montalentii	NT
Discoglossus sardus	LC
Discoglossus pictus*	NA

SPECIES	RED-LIST CATEGORY
D. sardus Hyères Islands	VU D2
D. sardus Corsica	LC
Bombina variegata	VU B2ab(i,ii,iii,iv,v)
Pelobates cultripes	VU B1ab(i,ii,iii,iv)
Pelobates fuscus	EN B1ab(i,ii,iii,iv)+2ab(i,ii,iii,iv)
Pelodytes punctatus	LC
Bufo bufo	LC
Bufo calamita	LC
Bufo viridis	NT
B. viridis Alsace and Lorraine	EN B2ab(i,ii,ii)
B. viridis Corsica	LC
Hyla arborea	LC
Hyla meridionalis	LC
Hyla sarda	LC
Xenopus laevis*	NA
Lithobates catesbeiana*	NA
Pelophylax bedriagae*	NA
Pelophylax kl. esculentus	LC
Pelophylax kl. grafi	DD
Pelophylax lessonae	NT
Pelophylax perezi	NT
Pelophylax ridibundus	LC
Rana arvalis	CR B2ab(i,ii,iii,iv,v)
Rana dalmatina	LC
Rana pyrenaica	EN B1ab(ii,iii,iv)+2ab(ii,iii,iv)
Rana temporaria	LC

V. Conclusions

France contains a great variety of biogeographical units. It is therefore more relevant to work at the regional or local scale when it comes to the conservation of amphibians. Indeed, conservation issues are different and, for example, are not treated in the Rhine valley in the same way as along the Mediterranean coast. It is therefore quite difficult to set up national programmes and assessments regarding the conservation of amphibians. Still, many topics are common among different regions and experiences can be shared.

The main threats to amphibians in France nowadays concern their habitats, both terrestrial and aquatic. Fragmentation and destruction of habitats, the drying and loss of wetlands, and the pollution of land and water induced by agricultural and industrial practices, are among the factors that have a negative impact on amphibian assemblages. All of these combined can lead to local declines, as is the case for rare species like the common spadefoot in Moselle, or previously widespread species such as the "green frogs" in the western bocage of Deux-Sèvres.

The most endangered species in France nowadays seems to be the moor frog, *Rana arvalis*. This species may have disappeared totally from the Rhine Valley and only survives now in two isolated localities in the Nord department. As with the moor frog, several other amphibian species reach the eastern limit of their ranges in France and have a conservation status that is not favourable;

they are therefore rather sensitive to disturbances in France. Also, there are other species with a very restricted global distributional range, especially in montane areas, and they are also subject to conservation projects.

During the past ten years, many naturalists became increasingly interested in the topic of amphibian conservation, and many associations were constituted in French regions and departments in order to raise awareness and to work on conservation issues for amphibians. Monitoring and conservation programmes such as action plans are implemented in many parts of France and, hopefully, an actual amphibian task force will be created in order to address more effectively the preservation of the very rich and diverse amphibian fauna.

VI. Acknowledgements

We would like to thank the people who helped us gather data for this article (in alphabetical order): Alexandre Boissinot, Matthieu Berroneau, Marc Cheylan, Pierre Olivier Cochard, Christophe Coïc, Pierre-André Crochet, Tony Dejean, Pascal Dubech, Rémi Duguet, Christophe Eggert, Philippe Geniez, Pierre Grillet, Olivier Grosselet, Robert Guyétant, Patrick Haffner, Florian Kirchner, Bernard Legarff, Jean Lescure, Jean-Christophe de Massary, Gilles Pottier, Dirk Schmeller, Jacques Thiriet, Jean-Marc Thirion.

VII. References

ACEMAV coll., Duguet, R. and Melki, F. (eds), 2003. *Les Amphibiens de France, Belgique et Luxembourg.* Biotope (collection Parthénope), Mèze. Andrenone, F., Miaud C., Bergo P. E., Doglio, S., Stocco, P., Ribéron, A. and Gautier, P., 2004. Living a at high altitude: testing the effect of life-history traits upon the conservation of *Salamandra lanzai* (Amphibia, Salamandridae). *Italian Journal of Zoology* **71** (suppl. 1): 35–43.

Andreone, F., Miaud, C., Bergò, P. E., Doglio, S., Stocco, P., Riberon, A. and Gautier, P., 2004. Living at high altitude: testing the effects of life history traits upon the conservation of Salamandra lanzai (Amphibia, Salamandridae). Italian Journal of Zoology **71**: 35–43.

Barnaud, G. and Fustec, E. 2007. *Conserver les Zones humides: Pourquoi? Comment?* Éducagri éditions/Quae éditions, Dijon/Versailles.

Berroneau, M., 2010. *Guide des Amphibiens et Reptiles d'Aquitaine.* Cistude Nature, Le Haillan.

Boissinot, A., 2010. *Amphibiens et Paysages bocagers: Influence de la Structure du Biotope de Reproduction et de la Configuration paysagère.* Éditions universitaires européennes, Sarrebrück, Germany.

Bosch J., Martinez-Solano, I. and Garcia-Paris, M. 2001. Evidence of a chytrid fungus infection involved in the decline of the common midwife toad (*Alytes obstetricans*) in protected areas of central Spain. *Biological Conservation* **97**: 331–337.

Castanet, J. and Guyétant, R. (coord.), 1989. *Atlas de Répartition des Amphibiens et Reptiles de France.* Société herpétologique de France, Paris.

Comité français de l'UICN, 2009. *La Liste rouge des Espèces menacées en France.* Reptiles et Amphibiens de France métropolitaine. Paris.

Cruz, M. J. and Rebelo, R., 2007. Colonization of freshwater habitats by an introduced **crayfish**, *Procambarus clarkii*, in Southwest Iberian Peninsula. *Hydrobiologia* **575**: 191–201.

Cruz, M. J., Rebelo, R. and Crespo, E. G., 2006. Effects of an introduced crayfish, *Procambarus clarkii*, on the distribution of south-western Iberian amphibians in their breeding habitats. *Ecography* **29**: 329–338.

Dejean, T., Miaud, C. and Ouellet, M., 2010a. La chytridiomycose: une maladie émergente des amphibiens. *Bulletin de la Société herpétologique de France* **134**: 27–46.

Dejean, T., Miaud, C. and Schmeller, D., 2010b. Protocole d'hygiène pour limiter la dissémination de la Chytridiomycose lors d'interventions de terrain. *Bulletin de la Société herpétologique de France* **134**: 47–50.

Detaint, M. and Coïc, C., 2005. La Grenouille taureau *Rana catesbeiana* dans le sud-ouest de la France. Premiers résultats du programme de lutte. *Bulletin de la Société Herpétologique de France* **117**: 41–56.

Eggert, C., 2002. Le déclin du Pelobate brun (*Pelobates fuscus*, Amphibien Anoures): De la biologie des populations à la structuration génétique. *Bulletin de la Société Zoologique de France* **127**: 273–279.

Ficetola ,G. F., Coïc, C., Detaint, M., Berroneau, M., Lorvelec, O. and Miaud, C., 2007. Pattern of distribution of the American bullfrog *Rana catesbeiana* in Europe. *Biological Invasions* **9**: 767–772.

Fouquet, A. and Measey, G., 2006. Plotting the course of an African clawed frog invasion in Western France. *Animal Biology* **56:** 95–102.

Gosá Oteiza, A., Rubio Pilarte, X. and Iraila Apaolaza, A., 2010. *Rana pyrenaica. Une relique des Pyrénées.* Aranzadi, Donostia-San Sebastián.

Holsbeeek, G., Mergeay, J., Holtz, H., Plötner, J., Volckaert, F. A. M. and De Meester, L., 2008. A cryptic invasion within an invasion and widespread introgression in the European water frog complex: consequences of uncontrolled commercial trade and weak international legislation. *Molecular Ecology* **17**: 5023–5035.

Joly, P., Miaud, C., Lehmann, A. and Grolet, O., 2001. Habitat matrix effect on pond occupancy in newts. *Conservation Biology* **15**: 239–248.

Knapp R. A. and Matthews, K. R., 2000. Non-native fish introductions and the decline of the mountain yellow-legged frog from within protected areas, *Conservation Biology* **14**: 428–438.

Lescure, J., 1984. La répartition passée et actuelle des pélobates (Amphibiens, Anoures) en France. *Bulletin de la Société Herpétologique de France* **29**: 45–59.

Lescure, J., Pichenot, J. and Cochard, P.-O., 2011. Régression de *Bombina variegata* (Linné, 1758) en France par l'analyse de sa répartition passée et présente. *Bulletin de la Société Herpétologique de France* **137**: 5–41.

Lescure, J. (coord.), Atlas des Amphibiens et Reptiles de France. MNHN/SHF, Paris. In press.

Llamas, A., Martinez-Gil, O. and Arribas, O. J., 1998. *Rana pyrenaica*, a new species for the French herpetofauna. *Boletín de la Asociación Herpetológica Española* **9**: 12–13.

Orizaola, G. and Braña, F., 2006. Effect of salmonid introduction and other environmental characteristics on amphibian distribution and abundance in mountain lakes of northern Spain. *Animal Conservation* **9**: 171–178.

Pointerau, P., 2002. Les haies, évolution du linéaire en France depuis 40 ans. *Le Courrier de l'Environnement* **46**: 69–71.

Peeler, E. J., Oidtmann, B. C., Midtlyng, P. J., Miossec, L. and Gozlan, R. E., 2011. Non-native aquatic animals introductions have driven disease emergence in Europe. *Biological Invasions* **13**: 1291–1303.

Poiret, M., 2005. Évolution des cultures et impacts sur l'environnement. Unpublished report.

Ribéron ,A., Desmet, J.-F. and Miaud, C., 2003. Validation génétique de la présence de la Salamandre noire *Salamandra atra* en France (Département de la Haute-Savoie). *Bulletin de la Société Herpétologique de France* **106**: 4–9.

Schmeller, D. S., Pagano, A., Plénet, S. and Veith, M., 2007. Introducing water frogs – Is there a risk for indigenous species in France? *Comptes Rendus Biologies* **330**: 684–690.

Semlitsch, R. (ed.), 2003. *Amphibian Conservation*. Smithsonian Institution Press. Washington DC.

Speybroeck, J., Beukema, W. and Cochet, P.-A., 2010. A tentative species list of the European herpetofauna (Amphibia and Reptilia) - an update. *Zootaxa* **2492**: 1–27.

Stöck, M., Sicilia, A., Belfiore, N. M., Buckley, D., Lo Brutto, S., Lo Valvo, M. and Arculeo, M., 2008. Post-Messinian evolutionary relationships across the Sicilian channel: Mitochondrial and nuclear markers link a new green toad from Sicily to African relatives. *BMC Evolutionary Biology* **8**: 56–74.

Stuart, S., Hoffmann, M., Chanson, J., Cox, N., Berridge, R., Ramani, P. and Young, B. (eds), 2008. *Threatened Amphibians of the World*. Lynx Edicions, Barcelona, Spain; IUCN, Gland, Switzerland; Conservation International, Arlington, Virginia, USA.

Thirion, J.-M., 2006. Le Pélobate cultripède *Pelobates cultripes* (Cuvier, 1829) sur la façade atlantique française : chorologie, écologie et conservation. Diplôme de l'Ecole Pratique des Hautes Etudes, Laboratoire de Biogéographie et Ecologie des Vertébrés, Montpellier.

Vacher, J.-P., 2010a. La grenouille des champs. Pp. 142–148 in *Atlas de Répartition des Amphibiens et Reptiles d'Alsace*, ed by J. Thiriet and J.-P. Vacher (coordinator). Bufo, Colmar/Strasbourg.

Vacher, J.-P., 2010b. Statut actuel de la grenouille des champs *Rana arvalis* Nilsson, 1842 (Amphibia: Ranidae) sur la bande rhénane en Alsace. *Bulletin de la Société d'Histoire naturelle et d'Ethnographie de Colmar* **69**: 3–12.

Vacher, J.-P. and Dutilleux, G., 2010. Le pélobate brun. Pp.82–90 in *Atlas de Répartition des Amphibiens et Reptiles d'Alsace*, coordinated by J. Thiriet and J.-P. Vacher. Bufo, Colmar/Strasbourg.

Vacher, J.-P., Falguier, A., Pinston, H. and Craney, E. 2008. The moorfrog *Rana arvalis* Nilsson, 1842 in eastern France. Distribution, ecology and conservation. Pp. 283–290 *in Der Moorfrosch/The Moor Frog*, coord. by D. Glandt. *Zeitschrift für Feldherpetologie*, Supplement 13.

37 Conservation and declines of amphibians in Spain

Cesar Ayres, Enrique Ayllon, Jaime Bosch, Alberto Montori, Manuel Ortiz-Santaliestra, and Vicente Sancho

Abbreviations and acronyms used in the text and references:

AHE	*Herpetological Association of Spain (Asociación Herpetológica Española)*
IREC	*Institute for Research into Hunting Resources*
IUCN	*International Union for the Conservation of Nature*
MNCN	*National Museum of Natural Sciences (Museo Nacional de Ciencias Naturales)*
SARE	*Monitoring of Amphibians and Reptiles of Spain*
SIARE	*Information Server of Spanish Amphibians and Reptiles*
UTM	*Universal Transverse Mercator*

I. Introduction

The geographic location of the Iberian Peninsula makes it a bridge between the Euro-Siberian and North-African herpetofauna. There are two climatic zones present in Spain. The Atlantic climate, restricted to a 100 km band along the northern coast, is wet and cool and the Mediterranean climate is typical of most of the Spanish territories, with warm and dry summers, and cool and wet winters.

These climatic areas make the Iberian Peninsula one of the richest herpetofaunal areas of Europe. There are 30 amphibian species (20 anurans and 10 caudates) reported from continental and insular Spain, and four additional species in the North-African territories. There is another anuran species reported from the northeastern coast of Spain, the Mediterranean painted frog (*Discoglossus pictus*), established from captive populations (Llorente *et al.* 2002).

The Northern Atlantic climatic zone probably acted as a refuge during the Quaternary (Vargas and Real 1997). This is one of the reasons there is such a high proportion of endemic species in the Iberian Peninsula (25%). On the other hand, the Strait of Gibraltar constrained northward movement of the African herpetofauna, thus rendering the Iberian herpetofauna more similar to that of northwestern Europe than that of northern Africa (Oosterbroek and Arntzen, 1992).

Human activities have influenced amphibian distribution and richness throughout history. Many species have been introduced, especially to islands during 8,000 years of sea traffic around the Mediterranean and along the coasts of the continents (Mayol 1997).

II. Declining species

There are several factors affecting amphibians in Spain: habitat loss and degradation, pollution, alien species, and diseases. The last factor has high relevance in recent years, causing declines in populations of the midwife toad (*Alytes obstetricans*) (Bosch 2001). This was the first episode of infection by *Batrachochytrium dendrobatidis* reported in Europe. Later, the chytrid fungus affected common toad (*Bufo bufo*) and fire salamander (*Salamandra salamandra*) populations in central Spain (Bosch and Martinez-Solano 2006). Probably Spain is the European country most impacted by chytridiomycosis (Garner *et al.* 2005). This disease even affected recovery programs of endangered species, as was the case for the captive breeding programme of the Mallorcan midwife toad (*Alytes muletensis*) (Walker *et al.* 2008). Also, some episodes of mass mortalities in newts related to attack by iridovirus have been reported (Balseiro *et al.* 2010; Soares *et al.* 2003).

Chytrid fungi may have a greater impact due to the rise of temperatures linked with climatic change (Bosch *et al.* 2007). Also, climatic change can pose a threat for Iberian amphibians in a scenario in which ranges contract under hydrological stress (Araujo *et al.* 2006). Taking in account the "no dispersal" assumption, it seems possible that many amphibian species will lose suitable climatic areas by 2050. Spain will be probably one of the most affected countries (Henle *et al.* 2008).

Alien species have been proposed as a cause of decline in many parts of the Iberian Peninsula (Cruz and Rebelo 2005); these are associated with important economic costs (Vila *et al.* 2010). Crayfish caused collapse of amphibian assemblages in some areas of the Iberian Peninsula (Cruz *et al.* 2008) and excluded some species from breeding ponds (Cruz *et al.* 2006). Tadpoles are not able to recognize chemical cues emanating from crayfish (Gomez-Mestre and Diaz-Paniagua 2011) or alien turtles (Polo-Cavia *et al.* 2010). Alien fishes also have direct effects on amphibian populations and indirect effects on the distribution of amphibians (Bosch *et al.* 2006, Orizaola and Braña 2006).

Environmental pollution is also among the major threats to amphibian populations. Because of the combination of terrestrial and aquatic phases in their life cycles, amphibians are exposed to multiple contaminants throughout their entire lives. Most studies with Spanish amphibians focus on the aquatic stages (Ortiz-Santaliestra *et al.* 2006), although the accumulation of agrochemicals (i.e., pesticides and fertilizers) on crop fields has been shown as potentially harmful for adult amphibians during their migration (Ortiz-Santaliestra *et al.* 2005). The absence of field monitoring makes it difficult to establish cause-effect relationships in this context; however, the few available data indicate that agriculturally related pollution is an important detriment to amphibian survival. Egea-Serrano *et al.* (2009) demonstrated that environmental nitrogen levels occurring in the Segura River Basin reduced survival of larval green frogs (*Pelophylax perezi*).

III. Conservation measures and monitoring programmes

Currently there is no National Action Plan for any amphibian species in Spain. Many regional governments, however, have implemented Regional Action plans to recover endangered species. Probably the most famous conservation program was the recovery of the Mallorcan Midwife toad (Griffiths and Kuzmin 2012) in cooperation with the captive breeding program of the Durrell and Barcelona zoos. Another conservation program is the Peñalara Breeding Centre, where there is a captive breeding program of midwife toads (*Alytes obstetricans*) and Iberian brown frogs (*Rana iberica*), aimed at recovering populations affected by chytridiomycosis and alien species. There is a captive breeding program for the reinforcement of the populations of the Montseny brook newt (*Calotriton arnoldi*), coordinated by the *Generalitat de Catalunya*. This regional government has also initiated a recovery project for some endangered populations of the ribbed newt (*Pleurodeles waltl*).

In the same area the Herpetological Association of Spain (AHE) is coordinating a project for the reintroduction of *A. obstetricans* and the Natterjack toad (*Bufo calamita*). In the Valencia region a LIFE project was conducted between 2006 and 2009 to recover and create habitats for amphibians. One hundred ponds have been created and/or restored, 20 faunal reserves have been declared to protect important habitat for *P. waltl*, the Iberian painted frog (*Discoglossus jeanneae*) and the spadefoot toad (*Pelobates cultripes*), among other species.

Recently, the AHE has developed a herpetological information server (SIARE), (http://siare. herpetologica.es) which is a system for detecting and monitoring the loss of biodiversity in the Spanish herpetofauna.

The SIARE has three complementary applications. The first is a database of herpetological data with about 200,000 records of amphibians and reptiles from the Spanish territory. This application is available online and provides access to precise distributional data and the option of supporting it by clicking on the map of Spain. These data are accumulated in an observer's database and, in turn, in the database of general data, once accepted. Users can download their particular database in Excel format at any time. It functions, therefore as a personal database accessible from any computer connected to the Internet, and at the same time enables users to participate in a comprehensive national project. The data that users enter in the database are displayed geographically in the form of presence within the 10 x10 km grid of the UTM projection, and is downloadable only by the user. Users always are allowed to ask for additional data on the presence of amphibians with a precision of 10 x10 km for any study (see application for requesting data in the website of the AHE, www.herpetologica.org). The purpose of this application is to detect changes in the distribution of species over time.

Another application available on-line is the AHEnuario. This is an application similar to the above, but intended for entering phenological observations (referring to clutches, songs, larvae, courtship, etc.). Like the previous database, this application can receive and manage observations of amphibians and reptiles from each user while allowing export for inclusion in the general database. The purpose of this application is to detect possible changes in species' phenology.

Finally, the SARE (Monitoring of Amphibians and Reptiles of Spain) is a volunteer program that seeks to involve all naturalists, technicians, biologists, and managers in the long-term monitoring of populations of amphibians and reptiles to establish long series of data so as to determine the evolution of populations. At the same time, it will determine the most reliable indicators for detecting decreases or other changes in amphibian assemblages.

There are other amphibian monitoring programs in protected places of Spain, such as the programs running in National Parks (Doñana, Cabañeros, Picos de Europa), or Natural Parks (Peñalara, Montes de Valsain, Parque del Sureste).

IV. Red List

In February 2011 a new Red List and legislation was approved by the Spanish government. This catalogue of endangered species doesn't use the IUCN categories, and it has two different lists: the Red list, with the species catalogued as Endangered or Vulnerable, and a list of species included in the catalogue which deserve protection but with no specific category assigned.

The Red List for 2011 included *Alytes muletensis* and *Calotriton arnoldi* as EN (Endangered) and *Salamandra algira, Chioglossa lusitanica, Messotriton alpestris, Alytes dickhilleni, Rana dalmatina,* and *Rana pyrenaica* as VU (Vulnerable). The following species were considered of special concern: *Calotriton asper, Lissotriton boscai, Lissotriton helveticus, Pleurodeles waltl, Triturus marmoratus, Triturus pygmaeus, Alytes obstetricans, Alytes cisternasii, Discoglossus galganoi, Discoglossus jeannae, Disco-*

glossus pictus, Rana iberica, Rana temporaria, Pelobates cultripes, Pelodytes ibericus, Pelodytes puntactus, Hyla arborea, Hyla meridionalis, Bufo calamita, Bufo balearicus.

This change in the Red List represents a step backwards in the conservation of Spanish amphibians. Some species should be under a higher category of risk, and others that are not now included in the catalogue should be.

Salamandra salamandra, Bufo bufo, and some species from the North African territories remain excluded from the list but, controversially, one introduced species (*D. pictus*) is included as a species of special concern. This species, which has been expanding its range since its introduction in 1906, seems to compete during its larval stage with larval *B. calamita*.

V. Conclusions

Spain possesses a rich amphibian diversity that has an uncertain future. The changes in habitat caused by human activities, as well as the potential effects of climatic change could have a major impact on amphibian populations, especially in the southern part of the country. These factors could interact synergistically with other stressors, such as pathogens (*Saprolegnia*, iridovirus, chytrid fungi), causing severe declines in some populations.

VI. Acknowledgments

We are grateful to the following persons for their assistance:

Compiler: Cesar Ayres (AHE-Conservation Issues)

Contributors: Enrique Ayllon (AHE), Jaime Bosch (MNCN), Albert Montori (Departamento de Biología Animal, Universidad de Barcelona), Manuel Ortiz-Santaliestra (IREC), Vicente Sancho (Coordinator Project LIFE-Anfibios Valencia).

VII. References

Araújo, M. B., Thuiller, W. and Pearson, R. G., 2006. Climate warming and the decline of amphibians and reptiles in Europe. Journal of Biogeography **33**: 1712–1728.

Balseiro, A., Dalton, K. P., del Cerro, A., Márquez, I., Parra, J. M., Prieto, J. M. and Casais, R., 2010. Outbreak of common midwife toad virus in alpine newts (Mesotriton alpestris cyreni) and common midwife toads (Alytes obstetricans) in Northern Spain: A comparative pathological study of an emerging ranavirus. The Veterinary Journal **186**: 256–258.

Bosch, J., Martínez-Solano, I. and García-París, M., 2001. Evidence of a chytrid fungus infection involved in the decline of the common midwife toad (*Alytes obstetricans*) in protected areas of central Spain. *Biological Conservation* **97**: 331–337.

Bosch J. and Martínez-Solano, I., 2006. Chytrid fungus infection related to unusual mortalities of *Salamandra salamandra* and *Bufo bufo* in the Peñalara Natural Park (Central Spain). *Oryx* **40**: 84–89

Bosch, J., Carrascal, L. and Fisher, C., 2007. Climate change and outbreaks of amphibian chytridiomycosis in a montane area of Central Spain; is there a link? *Proceedings of the Royal Society of London* **B274**: 253–260.

Cruz, M. J., Rebelo, R., 2005. Vulnerability of Southwest Iberian amphibians to an introduced crayfish, *Procambarus clarkii. Amphibia-Reptilia* **26**: 293–303.

Cruz, M. J., Pascoal, S., Tejedo, M. and Rebelo, R., 2006. Predation by an exotic crayfish, *Procambarus clarkii,* on Natterjack Toad, *Bufo calamita,* embryos: Its role on the exclusion of this amphibian from its breeding ponds. *Copeia* **2006**: 274–280.

Cruz, M. J., Segurado, P., Sousa, M. and Rebelo, R., 2008. Collapse of the amphibian community of the Paul do Boquilobo Natural Reserve (central Portugal) after the arrival of the exotic

Egea-Serrano, A., Tejedo, M. and Torralba, M., 2009. Populational divergence in the impact of three nitrogenous compounds and their combination on larvae of the frog *Pelophylax perezi* (Seoane, 1885). *Chemosphere* **76**: 869–877.

Garner, T. W. J., Walker, S., Bosch, J., Hyatt, A. D., Cunningham, A. A. and Fisher, M. C., 2005. Widespread European distribution of a global amphibian pathogen. *Emerging Infectious Disease* **11**: 1639–1641.

Gomez-Mestre, I. and Díaz-Paniagua, C., 2011. Invasive predatory crayfish do not trigger inducible defences in tadpoles. *Proceedings of the Royal Society B* 278: 3364–3370.

Griffiths, R. A and Kuzmin, S. L., 2011. Captive breeding of amphibians for conservation. Pp. 3687–3703 in *Conservation and Decline of Amphibians: Ecological Aspects, Effects of Humans, and Management*, ed by H. Heatwole and J. W. Wilkinson, vol. 10 in the series *Amphibian Biology*, ed by H. Heatwole. Surrey Beatty & Sons, Baulkham Hills.

Henle K., Dick D., Harpke A., Kühn I., Schweiger O., and Settele, J., 2008. Climate Change Impacts on European Amphibians and Reptiles. Proceedings of the Convention on the Conservation of European Wildlife and Natural Habitats, Standing Committee, 28th meeting, Strasbourg, 24–27 November 2008.

Llorente, G. A., Montori, A., Santos, X., and Carretero, M. A., 2002. *Discoglossus pictus*. Pp. 91–93 in *Atlas y Libro Rojo de los Anfibios y Reptiles de España*, ed by J. M. Pleguezuelos, R. Márquez, and M. Lizana. Dirección General de la Naturaleza - Asociación Herpetológica Española (Second Impression), Madrid.

Mayol, J., 1997. Biogeografía de los anfibios y reptiles de las islas Baleares. Pp. 371–379 in *Distribución y Biogeografía de los Anfibios y Reptiles en España y Portugal*, ed by J. M. Pleguezuelos. *Monografías de Herpetología* **3**. Universidad de Granada, Asociación Herpetológica Española.

Oosterbroek, P. and Arntzen, J. W., 1992. Area-cladograms of circumedoterranean taxa in relation to Mediterranean palaeogeography. *Journal of Biogeography* **19**: 3–20.

Orizaula, G. and Braña, F., 2006. Effect of salmonid introduction and other environmental characteristics on amphibian distribution and abundance in mountain lakes of northern Spain. Animal Conservation **9**: 171–178.

Ortiz-Santaliestra, M. E., Marco, A. and Lizana, M., 2005. Sensitivity and behavior of the Iberian newt, *Triturus boscai*, under terrestrial exposure to ammonium nitrate. *Bulletin of Environmental Contamination and Toxicology* **75**: 662–669.

Ortiz-Santaliestra, M. E., Marco, A., Fernandez, M. .J. and Lizana, M., 2006. Influence of developmental stage on sensitivity to ammonium nitrate of aquatic stages of amphibians. *Environmental Toxicology and Chemistry* **25**: 105–111.

Polo-Cavia, N., Gonzalo, A., López, P. and Martín, J., 2010. Predator recognition of native but not invasive turtle predators by naïve anuran tadpoles. Animal Behaviour **80**: 461–466.

Soares, C., Alves de Matos, A. P., Arntzen, J. W., Carretero, M. and Loureiro, A., 2002. Amphibian mortality in a National Park in the North of Portugal. *Froglog* 56: 1.

Vargas, J. M., Real, R. and Guerrero, J. C., 1998. Biogeographical regions of the Iberian Peninsula based on freshwater fish and amphibian distributions. *Ecography* **21**: 371–382.

Vilà, M., Basnou, C., Pyšek, P., Josefsson, M., Genovesi, P., Gollasch, S., Nentwig, W., Olenin, S., Roques, A., Roy, D., Hulme, P. E. and DAISIE partners, 2010. How well do we understand the impacts of alien species on ecosystem services? A pan-European, cross-taxa assessment. Frontiers in Ecology and the Environment 8:135–144.

Walker, S. F., Bosch, J., James, T. Y., Litvintseva, A. P., Oliver, J. A., Piña, S., García, G., Rosa, A. G., Cunningham, A. A., Hole, S., Griffiths, R. and Fisher, M. C., 2008. Invasive pathogens threaten species recovery programs. *Current Biology* **18**: R853–R854.

38 Conservation and declines of amphibians in Portugal

Rui Rebelo, Maria José Domingues Castro, Maria João Cruz, José Miguel Oliveira, José Teixeira, and Eduardo Crespo

I. Introduction – the country and its amphibian fauna

II. Threats to amphibians

III. Declining species or species of special concern for conservation

IV. Conservation measures and monitoring programmes

V. Conclusions

VI. Acknowledgments

VII. References

Abbreviations and acronyms used in the text and list of references:

ASL	*Above sea level*
CIBIO	*Research Centre in Biodiversity and Genetic Resources*
CIBIO-Div	*Science Divulgation Division of CIBIO*
CHARCOScomBIO	*Research, Conservation and Outreach of the Biodiversity of Temporary Ponds*
CSIC	*Consejo Superior de Investigaciones Científicas (Spanish National Reasearch Council)*
EDP	*Energias de Portugal*
ICNB	*National Institute for the Conservation of Nature and Biodiversity*
IUCN	*International Union for the Conservation of Nature*
LIFE	*The European Union's funding instrument for the environment*
SIAM	*Project Climate Change in Portugal. Scenarios, Impacts and Adaptation Measures*
SNPRCN	*Serviço Nacional de Parques, Reservas e Conservação da Natureza (former name of the current governmental Institute for the Conservation of Nature and Biodiversity.*

I. Introduction – the country and its amphibian fauna

The area of continental Portugal is 91,951 km² and the country is roughly rectangular in shape, with a main north-south axis 560 km long. It is almost entirely located between the parallels 37º N and 42º N, within the range of Mediterranean climate, but with a long Atlantic coast (Fig. 38.1). Orography and rainfall ultimately divide the country into two main biomes. The region to the north of the Tejo River includes 95% of all the mountain ranges above 400 m asl (Fig. 38.1), including the highest mountain systems (the highest is Serra da Estrela, with a summit at 1,993 m). South of the Tejo River, the land is mostly flat, with a few, isolated mountain ranges. Rainfall is also highly heterogeneous, with a marked reduction from north to south and from west to east. While in the Northwest Mountains annual precipitation may reach 3,000 mm, one of the highest records for the whole of Europe, in the southeastern semi-arid plains it may fall below 400 mm. The country may thus be roughly divided in two parts along a southwest/northeast line located at the northern limit of the Tejo Valley: the mountainous Northwest, with an Atlantic climate, and the southeastern plain, with a typical Mediterranean climate. The natural vegetation of these two regions differs accordingly, the Northwest being originally covered by deciduous oak forests and the Southeast by shrublands and open woodlands of evergreen oaks. The two major southern mountains – São Mamede (1,025 m) and Monchique (900 m) – are ecological islands with an Atlantic climate. These mountains are important for some amphibians.

The Portuguese amphibian fauna occurs in two distinct assemblages that occupy different bioclimatic regions (Carvalho *et al.* 2011). While species such as *Chioglossa lusitanica, Lissotriton helveticus,* and *Rana iberica* have affinities with the Atlantic climate and are established in the Northwest, the Southeast is occupied by species such as *Pleurodeles waltl, Alytes cisternasii,* and *Pelobates cultripes* that have affinities with a Mediterranean climate. Moreover, there are four cases of species-pairs in which each member of a pair occupies a different climatic zone; these are *Triturus marmoratus/T. pygmaeus, Alytes obstetricans/A. cisternasii, Pelodytes cf. punctatus/P. ibericus,* and *Hyla arborea/H. meridionalis* (the species with Atlantic affinities are mentioned first).

The history of Portuguese herpetology was thoroughly revised by Crespo (2008). The systematic recording of amphibian distribution in Portugal began with Crespo (1971) and Malkmus (1979), culminating in the first atlas of amphibians and reptiles for Portugal (Crespo and Oliveira 1989). Recently, the systematization of knowledge about the distribution and status of amphibians in Portugal led to publication of the national Red Data Book of Vertebrates (Cabral *et al.* 2005), and the National Atlas of Amphibians and Reptiles (Loureiro *et al.* 2008). One of the main results of the new atlas was a generalized and marked increase in the apparent sizes of the distributional ranges of all amphibians, probably not because of an actual expansion of geographic ranges, but rather merely establishing records for areas from which data were formerly deficient.

The Portuguese amphibian fauna includes at least 19 native species: 12 anurans and 7 caudates (Table 38.1). Nine species (47%) are Iberian endemics or near endemics. The Portuguese territory encompasses about 30% or more of the distributional range of six species (*Chioglossa lusitanica, Triturus pygmaeus, Lissotriton boscai, Alytes cisternasii, Discoglossus galganoi,* and *Rana iberica*) and two subspecies (*Salamandra salamandra gallaica* and *Lissotriton helveticus sequeirai*). Furthermore, there are two endemic Portuguese subspecies, confined to relatively small mountain ranges – *S. salamandra crespoi* in Monchique Mountain and surrounding areas in the Southwest (Malkmus 1991) and *Chioglossa lusitanica* longipes in the Estrela, Açor and Lousã mountain ranges (Arntzen *et al.* 2007). The specific status of the populations of *Pelodytes* sp. living along the western coast is under discussion, and may comprise a form confined to Portuguese territory (Crespo *et al.* 2008).

At least two exotic amphibian species are established in Portugal: *Triturus carnifex* was introduced to the Azorean island of São Miguel (Silva *et al.* 1997) and, recently, an established population of *Xenopus laevis* was discovered in two streams near Lisbon (Rebelo *et al.* 2010). There are also ancient successful introductions of *Rana (Pelophylax) perezi* onto several Azorean islands, as well as onto Madeira, and of *Hyla meridionalis* onto Madeira. There are recent reports of an introduction of *S. salamandra* onto São Miguel (Fonseca, personal communication).

II. Threats to amphibians

Almost all the factors identified as contributing to amphibian declines worldwide also occur in Portugal. Of the 54 papers focusing on ecology, conservation, or genetics of Portuguese amphibian populations published since 2000 (Thomson Reuters Web of KnowledgeSM search, 25 May 2011), 24 mention present or potential threats to these populations. The introduction of exotic predatory species is the most frequently cited threat (in nine papers), followed by climatic change (seven

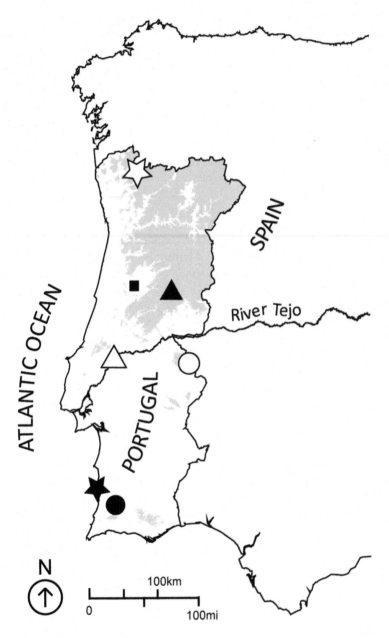

Fig. 38.1   Map of Portugal, showing its location on the western coast of the Iberian Peninsula and the course of the Tejo River. Areas above 400 m asl are shaded grey. Symbols show approximate locations of sites to which reference is made in the text. They are from north to south: White star = Peneda-Gerês National Park; Black square – the city of Coimbra; Black triangle = Serra da Estrela Natural Park; White triangle = Paul do Boquilobo Natural Reserve; White circle = San Mamede Natural Park; Black star = Sudoeste Alentejano and Costa Vicentina Natural Park; Black circle = Monchique mountain.

Table 38.1 Status of the native Portuguese amphibians, according to the IUCN (2010) Red List and to the Portuguese Red Data Book (Cabral *et al.* 2005). Nomenclature follows García-Paris *et al.* (2004).

VU – Vulnerable   **NT** – Near Threatened   **LC** - Least Concern   **N/A** – not assessed.
* Iberian endemic or near-endemic. Geographical/bioclimatic species affinities are indicated after the English species name: **NW** – Atlantic Northwest   **SE** – Mediterranean Southeast   **P** – all the territory.
"Trend" reflects our own opinion about the near-future prospects for each species: ↓ - decline: ↔ - stability.
Main threats: • - climatic change: † - invasive species: ‡ - emerging diseases.
Modifications to habitats are affecting all species.

SPECIES	IUCN	ICNB	Trend
Chioglossa lusitanica Barbosa du Bocage, 1864* Golden-striped salamander NW	VU	VU	↓•
Pleurodeles waltl Michahelles, 1830 Sharp-ribbed newt SE	NT	LC	↓•†
Salamandra salamandra (L., 1758) Fire salamander P	LC	LC	↓‡
Triturus marm5oratus (Latreille, 1800) Marbled newt NW	LC	LC	↓
Triturus pygmaeus (Wolterstorff, 1905)* Southern marbled newt SE	NT	N/A	↓•†
Lissotriton boscai (Lataste, 1879)* Bosca's newt P	LC	LC	↔
Lissotriton helveticus (Razoumowsky, 1789) Palmate newt NW	LC	VU	↓•†
Alytes obstetricans (Laurenti, 1768) Common midwife toad NW	LC	LC	↓•‡
Alytes cisternasii Boscá, 1879* Iberian midwife toad SE	NT	LC	↓•
Discoglossus galganoi Capula, Nascetti, Lanza, Bullini & Crespo, 1985* West Iberian painted frog P	LC	NT	↓•†
Pelobates cultripes (Cuvier, 1829)* Western spadefoot toad SE	NT	LC	↓•†
Pelodytes punctatus Bonaparte, 1838 Parsley frog P	LC	N/A	↓•†
Pelodytes ibericus Sánchez-Herráiz, Barbadillo, Machordom & Sanchiz, 2000* Iberian parsley frog SE	LC	N/A	↓•†
Bufo bufo (L., 1758) Common toad P	LC	LC	↓†‡
Bufo calamita Laurenti, 1768 Natterjack toad P	LC	LC	↔
Hyla arborea (L., 1758) Common treefrog NW	LC	LC	↔
Hyla meridionalis Boettger, 1874 Mediterranean treefrog SE	LC	LC	↔
Rana iberica Boulenger, 1879* Iberian frog NW	NT	LC	↓•
Rana (Pelophylax) perezi Seoane, 1885* Green frog P	LC	LC	↔

papers) and modification of habitats (six papers); environmental contamination (specifically uranium mining) is referenced in two papers.

Modification of habitats has accelerated, especially after the integration of Portugal into the European Union in 1986, and is mainly associated with changes in the traditional systems of agricultural production, including cases of land abandonment and shrub encroachment in marginal land, as well as an intensification of agriculture in productive soils (Pereira *et al.* 2004). Furthermore, there was a large expansion of the road network, construction of large dams, and urbanization of the coastal strip. Among the alterations with most impact on amphibian populations, the conversion of extensive areas to plantations of *Pinus pinaster* and *Eucalyptus* spp., as well as the construction of large dams, have been most frequent in the Northwest, while in the Southeast the main problem has been the drainage of temporary ponds, or their conversion to permanent reservoirs, usually as a result of the new agricultural practices.

Despite their dry summer, the southern plains are prone to flooding during the winter, thereby maintaining temporary ponds of different physiographies; these constitute the favoured breeding habitat for all the amphibians. However, as soon as intensive agricultural practices are implemented these valuable habitats are quickly eliminated, even inside protected areas. A recent study showed that 57% of the temporary ponds located in irrigated agricultural lands within the Sudoeste Alentejano and Costa Vicentina Natural Park, one of the largest (131,000 ha) protected areas in Portugal (Fig. 38.1), were filled and converted to intensive agriculture between 1991 and 2009 (Beja 2010). Ponds located in non-irrigated land also disappeared, but at a slower rate (13%). Beja and Alcazar (2003) showed that two of the factors that contributed to a reduction in the abundance of the nine amphibian species that breed in that region were intensification of agriculture and transformation of ponds into permanent reservoirs.

The expansion of exotic species in freshwater habitats may be even more destructive and difficult to counteract. Beja and Alcazar (2003) referred to the introduction of exotic predators (fish and crayfish) as one of the main factors contributing to the reduction of amphibian abundance in the Southwest. Virtually all of the dams and most of the water reservoirs have been stocked with exotic predatory fish, such as *Micropterus salmoides*, *Lepomis gibbosus*, and *Gambusia holbrooki* (Ribeiro *et al.* 2009). In the late 1970s the American red swamp crayfish *Procambarus clarkii* invaded Portugal from Spain. This species quickly expanded across the Mediterranean climatic region, where there were no native crayfish species, and reached very high densities (Correia 1995). The most impressive documented impact of an exotic species on amphibians in Portugal is the collapse of the amphibian assemblage of the Paul do Boquilobo Natural Reserve (a 550-ha wetland in the Tejo Valley) (Fig. 38.1), about ten years after the arrival of *P. clarkii* (Cruz *et al.* 2008). Eggs and larval stages of all the amphibians from the Southeast are vulnerable to predation by *P. clarkii* (Cruz and Rebelo 2005), and Cruz *et al.* (2006a,b) have shown that the presence of this crayfish is the most important negative predictor of reproductive success of most of these amphibians, with its highest impact on temporary ponds. Introductions of alien species continue, with the expansion of the red-eared slider *Trachemys scripta* (Loureiro *et al.* 2008) and the recent (2007) invasion by the catfish *Silurus glanis*, as well as by *Pacifastacus leniusculus*, a freshwater crayfish adapted to the Atlantic climate and which is spreading from Northern Spain into Portugal (Costa *et al.* 2009).

Teixeira and Arntzen (2002), Araújo *et al.* (2006), and Carvalho *et al.* (2010) showed that European amphibians will be particularly vulnerable to climatic change, due to their particular ecological requirements and their limited abilities to disperse. Such is the case for all Portuguese amphibians; especially those confined to mountain ranges or that are living at the southern limit of their distribution. The frequency of heat waves increased in Portugal during the past 20 years, particularly during spring (Santos and Miranda 2006), thereby accelerating the drying of ponds and hindering the reproduction of several species, especially in the South. This trend is expected to be accompanied by a general aridification of the country, with clear negative consequences for amphibians. However, conclusive evidence for amphibian declines that can be solely attributed to climatic changes is still lacking.

Emerging diseases are also becoming an issue for amphibian conservation in Portugal. Catastrophic mortality, associated with the development of severe skin lesions, of *T. marmoratus* and *L. boscai* in the high-elevation lakes (1,500 m asl) of Gerês National Park in the Northwest (Fig. 38.1) was attributed to an iridovirus (Froufe *et al.* 1999; Alves de Matos *et al.* 2004). Deaths, both of larvae and adults, were reported, beginning in the summer of 1998 and continuing today (C. Soares, personal communication). This disease might be non-native and is thought to have been transferred to the lakes by the introduced pumpkinseed sunfish, *L. gibbosus*. Finally, the chytrid fungus, *Batra-cochytrium dendrobatidis*, was recently found in Portugal (Garner *et al.* 2005), and at Serra da Estrela (Fig. 38.1) major outbreaks already have been found to affect *Alytes obstetricans*, a species that reproduces in lakes at high elevations (Rosa, PhD thesis in progress, beginning in 2011).

III. Declining species or species of special concern for conservation

The 2005 version of the Portuguese Red Data Book identified two Vulnerable species (*Chioglossa lusitanica* and *Lissotriton helveticus*) and one Near Threatened species (*Discoglossus galganoi*) (Table 38.1). Both caudates occur in the Atlantic Northwest, and their threatened status is related to their dependence on specific habitats (mountain brooks with pristine forest for the salamander and ponds for the newt). The palmate newt is clearly the most threatened amphibian in Portugal, with the smallest geographic range, most of which is located outside of protected areas.

Lissotriton helveticus and *D. galganoi* are dependent on traditional agricultural landscapes, using small fishless farm ponds (newt) and damp agricultural meadows (frog) to reproduce. Conversion of agricultural land to forest production altered large portions of their habitats; furthermore, the introduction of exotic fishes and especially of *P. clarkii* may lead to the extirpation of local popula-tions (Almeida *et al.* 2011).

The 2005 Portuguese Red Data Book did not assess the status of the different forms of *Pelodytes*, or of *Triturus pygmaeus*. The status of genetically differentiated populations of other species was also not assessed, such as those of S. *salamandra crespoi*, and the southwestern populations of *Lis-sotriton boscai*, whose specific status is questioned (Martínez-Solano *et al.* 2006). Some of these forms are currently living at the edge of their distributional range and consequently may be especially vulnerable to climatic changes.

The 2010 edition of the IUCN Red List of Threatened Species classified C. *lusitanica* as Vulner-able, and a further five species as Near Threatened (IUCN 2010) (Table 38.1). The Near Threatened species are Iberian endemics or near endemics adapted to Mediterranean climates, with the excep-tion of *Rana iberica*, a species endemic to northwestern Iberia. While there have been declines of this species in Spain (Bosch *et al.* 2006), in Portugal it is relatively common in suitable habitats. Nonetheless, pollution and habitat destruction have been implicated in local population declines, e.g., in the isolated population of São Mamede Mountain in the south (Balonas 2008) (Fig. 38.1).

The four remaining Near Threatened species (*P. waltl*, *T. pygmaeus*, *A. cisternasii*, and *P. cultripes*) are dependent on temporary ponds/streams for reproduction. Their declines have been associated with the introduction of exotic species (mainly *P. clarkii*) and with diversion of water for agriculture (Beja and Alcazar 2003; Cruz *et al.* 2006a; Nunes *et al.* 2010). A very strong decline of *P. waltl* and *T. pygmaeus* (as well as a local extirpation of *H. arborea* and *P.* cf. *punctatus*) was associated with the invasion by *P. clarkii* of a protected wetland (Cruz *et al.* 2008). This crayfish thrives in alluvial plains subject to a Mediterranean climate (Cruz and Rebelo 2007); its expansion will contribute to the fragmentation of amphibian populations. Other species that are dependent upon temporary ponds susceptible to invasion by *P. clarkii* are already experiencing population declines and in the future may be included in one of the threatened categories. The same will probably happen to *A. obstetricans* and possibly to *S. salamandra* and *B. bufo* in the Northwest, where *B. dendrobatidis* is currently spreading.

IV. Conservation measures and monitoring programmes

With the exception of *R. perezi* and of the exotic species, amphibians enjoy full legal protection in Portugal. During the past 20 years, public attitudes towards amphibians have changed and frogs and toads now have a high profile among children. Scientific knowledge of amphibian ecology, phylogeography, and conservation has also been steadily accumulating. However, there have been relatively few conservation measures directed specifically to this group.

During the past decade, two projects funded by the European Union LIFE program were directed toward conservation of amphibians, both concerning *Chioglossa lusitanica*. The results include a marked increase in knowledge about the biology, ecology, and population structure of this species, as well as habitat improvement at one of the few confirmed breeding sites of this cryptic species in Valongo Mountain, near the city of Porto (Teixeira *et al* 1998; Teixeira *et al.* 2001; http://www.cmvalongo.net/life/life.htm).

In 2009, the project CHARCOScomBIO – Research, Conservation and Outreach of the Biodiversity of Temporary Ponds, was proposed by a joint team from the universities of Porto and Lisbon; the aim of this project was to manage and protect temporary ponds across all the country. It was awarded a national conservation prize (the "EDP fund for Biodiversity"), which triggered the first conference on the Ecology and Conservation of Amphibians, in April 2010, and the creation of the first micro-reserves for amphibians in Portugal. The prize also supported the itinerant exhibition "Amphibians: a Paw on Water, Another on Land" (www.expoanfibios.org) and the National educational campaign "Ponds with Life" (www.charcoscomvida.org). The exhibition already has been displayed in seven cities, while the environmental education campaign encourages the inventory, adoption, construction, conservation, and pedagogical exploration of ponds. The exhibition and the National Educational Campaign were organized and managed by CIBIO-Div, a biodiversity conservation and science communication unit of CIBIO (Research Centre in Biodiversity and Genetic Resources), whose aim is to contribute to public awareness of amphibians and their fragile habitats.

Probably the most publicized conservation action ever conducted for a single amphibian species was the temporary translocation of an *Alytes obstetricans* population living around a century-old football field under renewal at the city of Coimbra (Fig. 38.1) (Castro *et al.* 2005a,b). This project was awarded the "National Environmental and Conservation Grant" by the Ford Motor Company and lasted from 2001 to 2008. A total of 312 adults (of the estimated 400–600 adults in the population) and 518 tadpoles were translocated from their habitat to a contiguous artificial habitat. The *ex-situ* artificial habitat was confined to prevent the toads from homing or escaping to nearby roads, and also to facilitate an accurate monitoring of the population. The midwife toads' population remained *ex-situ* for three years (2003–2006), after which 141 adults and 67 tadpoles were brought back to their original location, which had been ecologically restored (Castro *et al.* 2005a,b). Recent monitoring (2010/2011) showed an occupation of all the refugia by the toads similar to that before translocation. There was similar yearly activity patterns (calling activity and intensity), and similar use of the bodies of water for reproduction (MJDC and JMO, unpublished data). All the aquatic habitats that were created were adopted as breeding sites. This is the only amphibian translocation project in Portugal that has been subjected to a monitoring program sufficiently long to allow confirmation of its success.

As most of the systematic scientific work was initiated only during the past two decades, there is only one temporal series of population estimates – the one by Rebelo and Leclair (2003) for a population of *S. salamandra gallaica* in the Natural Park of Sintra-Cascais, 20 km northwest of Lisbon. This study lasted from 1990 to 1997 and indicated an apparently stable population.

Presently, amphibian-monitoring programmes are being carried out in the Peneda-Gerês National Park and in the temporary ponds of the southwest. Also, a pond survey was begun at the *B. dendrobatidis*-affected sites in Serra da Estrela in 2011. Assessments of road casualties are also being carried out regularly (Carvalho and Mira 2011). An eradication program of the introduced population of *X. laevis* is in its second year, with potential beneficial effects for the conservation of native amphibians, as this species is an asymptomatic vector of *B. dendrobatidis*.

V. Conclusions

About 20% of Portuguese territory is presently under some form of environmental legal protection. It is doubtful that any amphibian species will go extinct in Portugal in the near future, as the large distributional range of almost all the species (except for the palmate newt, *L. helveticus*) provides good prospects that local populations will survive in at least part of their present ranges. It is also clear, however, that amphibians have declined recently, and that the declines are accelerating. The loss of sites for reproduction, namely ponds, will certainly continue, and climatic warming will contribute to contractions of ranges and to local extirpations of species with restricted distributional ranges or specific ecological requirements. Nevertheless, two further threats are of special concern for the next decade: (1) the ongoing invasions by crayfishes - *P. clarkii* is still expanding along the western coast and into the few un-invaded places in the south, and *P. leniusculus* is invading the northwestern mountains; (2) the future spread of *B. dendrobatidis* in the central mountain range and other regions of high elevation. Finally, there are numerous potential interactions among these threatening factors (e.g., an invasive predator will constitute a higher threat for a population confined to a few sites due to previous habitat destruction or climatic change), and their impacts on amphibians are difficult to predict and to assess.

VI. Acknowledgments

Our work on amphibian conservation was supported by the Ford Motor Company (MJDC and JMO), and by EDP – Energias de Portugal, Sociedade Anonima (JT and RR). P. Segurado drew Fig. 38.1.

VII. References

Almeida, E., Nunes, A., Andrade, P., Alves, S., Guerreiro, C. and Rebelo R., 2011. Antipredator responses of two anurans towards native and exotic predators. *Amphibia-Reptilia,* in press.

Alves de Matos, A. P., Soares, C. and Carretero, M. A., 2004. *Iridovirus-like particles in lesions of diseased* Triturus marmoratus *from Carris and Batateiro.* Symposium on the Decline of the Populations of Amphibians. Societat Catalana de Herpetologia, Universitat de Lleida.

Araújo, M. B., Thuiller, W. and Pearson, R. G., 2006. Climate warming and the decline of amphibians and reptiles in Europe. *Journal of Biogeography* 33: 1712–1728.

Arntzen, J., Groenenberg, D. J., Alexandrino, J., Ferrand, N. and Sequeira, F., 2007. Geographical variation in the golden-striped salamander, *Chioglossa lusitanica* Bocage, 1864 and the description of a newly recognized subspecies. *Journal of Natural History* 41: 925–936.

Balonas, D., 2008. *Bases para a Conservação da Rã-castanha-ibérica* (Rana iberica) *no Parque Natural de São Mamede.* M. Sc. Thesis, University of Évora.

Beja, P., 2010. Countdown 2010? Lagoas temporárias, anfíbios e insucesso da conservação em Rede Natura 2000. Conference "Ecologia e Conservação de Anfíbios", Naturlink and CIBIO.

Beja, P. and Alcazar, R., 2003. Conservation of Mediterranean temporary ponds under agricultural intensification: an evaluation using amphibians. *Biological Conservation* 114: 317–326.

Bosch, J., Rincon, P. A., Boyero, L. and Martínez-Solano, I., 2006. Effects of introduced salmonids on a montane population of Iberian frogs. *Conservation Biology* 20: 180–189.

Cabral, M. J. (coordinator), Almeida, J., Almeida, P. R., Delllinger, T., Ferrand de Almeida, N., Oliveira, M. E., Palmeirim, J. M., Queiroz, A. I., Rogado, L. and Santos-Reis, M. (eds), 2005. *Livro Vermelho dos Vertebrados de Portugal.* Instituto da Conservação da Natureza, Lisbon.

Carvalho, F. and Mira, A., 2011. Comparing annual vertebrate road kills over two time periods, 9 years apart: a case study in Mediterranean farmland. *European Journal of Wildlife Research* 57: 157–174.

Carvalho, S. B., Brito, J. C., Pressey, R. L., Crespo, E. and Possingham, H. P., 2010. Simulating the effects of using different types of species distribution data in reserve selection. *Biological Conservation* 143: 426–438.

Carvalho, S. B., Brito, J. C., Crespo, E. and Possingham, H. P., 2011. Incorporating evolutionary processes into conservation planning using species distribution data: a case study with the western Mediterranean herpetofauna. *Diversity and Distributions* 17: 408–421.

Castro, M. J., Oliveira, J. M. and Tari, A., 2005a. Conflicts between urban growth and species protection: Can midwife toads (*Alytes obstetricans*) resist the pressure? Pp. 126–129 in *Herpetologica Petropolitana*, ed by N. Ananjeva, and O. Tsinenko. Proceedings of the 12th Ordinary General Meeting of the Societas Europaea Herpetologica, August 12–16, 2003, St. Petersburg. *Russian Journal of Herpetology* 12 (supplement).

Castro, M. J., Caldeira Cabral, F. H., Costa, R. T., Lemos, A., Morgado, P., Veríssimo, C. and Oliveira, J. M., 2005b. The ecological restoration project to protect an endangered population of midwife toads (*Alytes obstetricans*): An overview of the contribute from engineering and architecture. 13th Ordinary General Meeting of Societas Europaea Herpetologica. Bonn. Abstracts.

Correia, A. M., 1995. Population dynamics of *Procambarus clarkii* (Crustacea: Decapoda) in Portugal. *Freshwater Crayfish* 8: 276–290.

Costa, A. M., Bruxelas, S., Bernardo, J. and Teixeira, A., 2009. Colonization of Rio Maçãs (North Portugal) by two exotic crayfish *Pacifastacus leniusculus* and *Procambarus clarkii*. World Conference on Biological Invasions and Ecosystem Functioning (Biolief). Candeias Artes Gráficas, Porto.

Crespo, E. G., 1971. Anfíbios de Portugal Continental das colecções do Museu Bocage. *Arquivos do Museu Bocage* **3**: 203–204.

Crespo, E. G., 2008. História da Herpetologia em Portugal. Pp. 18–53 in *Atlas dos Anfíbios e Répteis de Portugal*, ed by A. Loureiro, N. Ferrand de Almeida, M. A. Carretero, O. S. Paulo (coordinator). Instituto da Conservação da Natureza e da Biodiversidade, Lisbon.

Crespo, E. G. and Oliveira, M. E., 1989. *Atlas da Distribuição dos Anfíbios e Répteis de Portugal Continental*. SNPRCN, Lisbon.

Crespo, E. G., Márquez, R., Pargana, J. and Tejedo, M., 2008. *Pelodytes* spp. Pp. 112–115 in *Atlas dos Anfíbios e Répteis de Portugal*, ed by A. Loureiro, N. Ferrand de Almeida, M. A. Carretero, O. S. Paulo (coordinator). Instituto da Conservação da Natureza e da Biodiversidade, Lisbon.

Cruz, M. J. and Rebelo, R., 2005. Vulnerability of Southwest Iberian amphibians to an introduced crayfish, Procambarus clarkii. *Amphibia-Reptilia* **26**: 293–304.

Cruz, M. J., Rebelo, R. and Crespo E. G., 2006a. Effects of an introduced crayfish, *Procambarus clarkii*, on the distribution of South-western Iberian amphibians in their breeding habitats. *Ecography* **29**: 329–338.

Cruz, M. J., Pascoal, S., Tejedo, M. and Rebelo, R., 2006b. Predation by an exotic crayfish, *Procambarus clarkii*, on natterjack toad, *Bufo calamita*, embryos: its role on the exclusion of this amphibian from its breeding ponds. *Copeia* **2006**: 274–280.

Cruz, M. J. and Rebelo, R., 2007. Colonization of freshwater habitats by an introduced crayfish, *Procambarus clarkii*, in Southwest Iberian Peninsula. *Hydrobiologia* **575**: 191–201.

Cruz, M. J., Segurado, P., Sousa, M. and Rebelo, R., 2008. Collapse of the amphibian community of the Paul do Boquilobo Natural Reserve (central Portugal) after the arrival of the exotic American crayfish *Procambarus clarkii*. *Herpetological Journal* **18**: 197–204.

Froufe, E., Arntzen, J. W. and Loureiro, A., 1999. Dead newts in Peneda Gerês, Portugal. *Froglog* **31**: 1.

IUCN, 2010. IUCN Red List of Threatened Species. Version 2010.4. http://www.iucn-redlist.org>. Downloaded on 25 May 2011.

García-Paris, M., Montori, A. and Herrero, P., 2004. *Amphibia. Lissamphibia. Fauna Ibérica*, ed by M. A. Ramos, J. Alba, X. Bellés, J. Gosálbez, A. Guerra, E. Macpherson, J. Serrano and J. Templado. vol. 24. Museo Nacional de Ciencias Naturales. CSIC, Madrid.

Garner, T. W. J., Walker, S., Bosch, J., Hyatt, A. D., Cunningham, A. A. and Fisher, M. C., 2005. Chytrid fungus in Europe. *Emerging Infectious Diseases* **11**: 1639–1641.

Loureiro, A., Ferrand de Almeida, N., Carretero, M. A. and Paulo, O. S. (coordinator), 2008. "Atlas dos Anfíbios e Répteis de Portugal". Instituto da Conservação da Natureza e da Biodiversidade, Lisbon.

Malkmus, R., 1979. Beitrag zur vertikalen Verbreitung der Herpetofauna Portugals. *Boletim da Sociedade Portuguesa de Ciências Naturais* **19**: 125–145.

Malkmus, R., 1991. Einige Bemerkugen zum Feuersalamander Portugals (*Salamandra salamandra gallaica* - Komplex) (Amphibia, Urodela: Salamandridae). *Zoologische Abhandlugen* **46**: 165–190.

Martinez-Solano, I., Teixeira, J., Buckley, D. and Garcia-Paris, M., 2006. Mitochondrial DNA phylogeography of *Lissotriton boscai* (Caudata, Salamandridae): evidence for old, multiple refugia in an Iberian endemic. *Molecular Ecology* **15**: 3375–3388.

Nunes, A. L., Cruz, M. J., Tejedo, M., Laurila, A. and Rebelo, R., 2010. Nonlethal injury caused by an invasive alien predator and its consequences for an anuran tadpole. *Basic and Applied Ecology* **11**: 645–654.

Pereira, H. M, Domingos, T. and Vicente, L. (eds), 2004. *Portugal Millennium Ecosystem Assessment: State of the Assessment Report*. Centro de Biologia Ambiental, Faculdade de Ciências da Universidade de Lisboa. Available online at http://ecossistemas.org.

Rebelo, R. and Leclair, H., 2003. Site tenacity in the terrestrial salamandrid *Salamandra salamandra*. *Journal of Herpetology* **37**: 440–445.

Rebelo, R., Amaral, P., Bernardes, M., Oliveira, J., Pinheiro, P. and Leitão, D., 2010. *Xenopus laevis* (Daudin, 1802), a new exotic amphibian in Portugal. *Biological Invasions* **12**: 3383–3387.

Ribeiro, F., Collares-Pereira, M. J. and Moyle, P. B., 2009. Non-native fish in the fresh waters of Portugal, Azores and Madeira Islands: a growing threat to aquatic biodiversity. *Fisheries Management and Ecology* **16**: 255–264.

Santos, F. D. and Miranda, P., 2006. "Alterações Climáticas em Portugal". Cenários, Impactos e Medidas de Adaptação - Projecto SIAM II. Gradiva, Lisbon.

Silva, L., Elias, R., Machado, E., Macedo, A., Sosa, F., Rebelo, J. and Nunes, A., 1997. Comparative study of three *Triturus cristatus* (Amphibia: Salamandridae) populations from São Miguel island (Azores). *Boletim do Museu Municipal do Funchal* **49**: 89–98.

Teixeira, J., Sequeira, F., Alexandrino, J. and Ferrand, N., 1998. Bases para a conservação da Salamandra-lusitânica, *Chioglossa lusitanica*. Instituto da Conservação da Natureza, Lisbon.

Teixeira, J., Ferrand. N. and Arntzen, J. W., 2001. Biogeography of the golden-striped salamander (*Chioglossa lusitanica*): A field survey and spatial modelling approach. *Ecography* **24**: 618–624.

Teixeira, J. and Arntzen, J. W., 2002. Potential impact of climate warming on the distribution of the golden-striped salamander, *Chioglossa lusitanica*, in Iberian Peninsula. *Biodiversity and Conservation* **11**: 2167–2176.

CPSIA information can be obtained
at www.ICGtesting.com
Printed in the USA
BVOW07s1808300317
479595BV00006B/1/P